Making Thirteen Colonies

A HISTORY OF US

Oxford University Press

In these books you will find explorers, farmers, cowboys, heroes, villains, inventors, presidents, poets, pirates, artists, slaves, teachers, revolutionaries, priests, musicians—the girls and boys, men and women, who all became Americans....

Making Thirteen Colonies

Joy Hakim

Oxford University Press
New York

Oxford University Press
Oxford New York Toronto
Delhi Bombay Calcutta Madras Karachi
Kuala Lumpur Singapore Hong Kong Tokyo
Nairobi Dar es Salaam Cape Town
Melbourne Auckland Madrid

and associated companies in
Berlin Ibadan

Designer: Mervyn E. Clay

Produced by American Historical Publications

Published by Oxford University Press, Inc.
200 Madison Avenue, New York, New York 10016

Library of Congress Cataloging-in-Publication Data
Hakim, Joy.
Making Thirteen Colonies/ by Joy Hakim.
p. cm.—(A history of US: 2)
Includes bibliographical references and index.
Summary: Presents the history of the United States from the colonization of the New World through the middle of the eighteenth century
ISBN 0-19-507747-4 (lib. ed.)—ISBN 0-19-507748-2 (pbk.)
1. United States—History—Colonial period, ca. 1600-1775—Juvenile literature. [1. United States—History—Colonial period, ca. 1600-1775.] I. Title. II. Series: Hakim, Joy. History of US; 2.
E178.3.H22 1993 vol. 2
[E188]
973.2—dc20 93-7194
CIP
AC

3 5 7 9 8 6 4 2
Printed in the United States of America
on acid-free paper

THIS BOOK IS FOR MY GRANDDAUGHTER NATALIE MARIE JOHNSON

INDIAN

I don't know who this Indian is,
A bow within his hand,
But he is hiding by a tree
And watching white men land.
They may be gods—they may be fiends—
They certainly look rum.
He wonders who on earth they are
And why on earth they've come.

He knows his streams are full of fish,
His forests full of deer,
And his tribe is the mighty tribe
That all the others fear.
—And, when the French or English land,
The Spanish or the Dutch,
They'll tell him they're the mighty tribe
And no one else is much.

They'll kill his deer and net his fish
And clear away his wood
And frequently remark to him
They do it for his good.
Then he will scalp and he will shoot
And he will burn and slay
And break the treaties he has made
—And, children, so will they.

We won't go into all of that
For it's too long a story,
And some is brave and some is sad
And nearly all is gory.
But just remember this about
Our ancestors so dear:
They didn't find an empty land,
The Indians were here.

—STEPHEN VINCENT BENÉT

An animal which has a head like a sucking pig...hair like a badger... the tail like a rat, the paws like a monkey, which has a purse beneath its belly, where it produces its young and nourishes them.

—Le Moyne d'Iberville, writing in 1699 in astonishment at an opossum, an animal unknown in Europe

Some ten years ago being in Virginia, and taken prisoner by the power of Powhatan their chief King, I received from this great savage exceeding great courtesy, especially from his son Nantaquaua, the most manliest, comeliest, boldest spirit I ever saw in a savage, and his sister Pocahontas, the King's most dear and well-beloved daughter, being but a child of 12 or 13 years of age, whose compassionate pitiful heart, of my desperate state, gave me much cause to respect her....Jamestown she as frequently visited as her father's habitation; and during the time of two or three years, she next under God, was still the instrument to preserve this Colony from death, famine and utter confusion.

—John Smith, adventurer and co-founder of Jamestown, Virginia

And we have now with Horror seen the Discovery of such a Witchcraft! An Army of Devils is horribly broke in upon...our English Settlements.

—Cotton Mather, Puritan clergyman and author

Contents

PREFACE

Our Mixed-up Civilization

Carved 4,000 years ago, this is the portrait of a man who ruled in Sumer at around the time Abraham lived.

A long time ago—actually it was almost 4,000 years ago—in the city of Ur, there lived a man named Abraham. Ur was in a country that is now known as Iraq but was then called Sumer.

Now you may be asking why we are in ancient Sumer when this is a book about United States history. Well, hold on. Abraham will turn out to be important—to people all over the world—and to us in America.

Abraham lived in an interesting urban center. In Sumer, between the Tigris (TY-griss) and Euphrates (yoo-FRAY-teez) rivers, people had learned to read and write and to build and govern cities (which hadn't been done much before). But something must have been wrong, because one day Abraham decided to move.

Abraham moved with his whole family and his cattle and oxen. He traveled northwest, following a fingernail of green land called the Fertile Crescent. He went to a place known as Canaan (KAY-nun); he called it the land of Israel. There Abraham had two sons: Isaac and Ishmael. The descendants of those sons founded two great religions. From Isaac's children came Judaism and the Jewish people; from Ishmael's came Islam and the Muslims.

Abraham must have been a restless type, because he traveled on, to Egypt and back to Canaan. In Egypt he found a spectacular civilization, where people could also read and write. The Egyptians had built big cities and tall pyramids. They called their rulers *pharaohs* (FAIR-ohz), and they kept slaves—as did many people in those days. Some of Abraham's people became slaves. Like all enslaved people, they longed to be free. Help was on its way.

This is a preface (PREFF-iss), which means it is a message to the reader about the book. This preface tells you something about the long-ago ancient world, and that will help you understand some of the people who sailed across the ocean and came to America.

Ancient Sumer
is sometimes called Sumeria or Mesopotamia.

Miriam, the pharaoh's daughter, finds Moses. Miriam's clothes, and her castle, don't look very ancient Egyptian. When this illustration was made, in the 15th century, artists just drew people the way they looked in their own time.

One day an Egyptian princess found a baby boy floating in a basket among the bulrushes at the edge of the river Nile. The baby was Moses. He became a great leader and led the Jewish people out of Egypt. The trip was filled with danger and adventure. It took Moses and the Jews 40 years to get to their destination: Israel. Actually, Moses didn't quite make it. You can read the story of that flight to freedom in the Hebrew Bible, which is also called the Old Testament.

Take a look at the map on page 13. Do you see Israel and Egypt? Did you notice that they are next to the Mediterranean Sea? Back in the old, old days, before cars, buses, trains, and airplanes, it was very hard traveling over land. It was much easier to get in a boat and sail away. So people who lived near a sea got around more than inland people. People living around the Mediterranean Sea traveled and traded ideas.

Greece borders the Mediterranean. More than 2,500 years ago, some very interesting people—artists, playwrights, scientists, and philosophers—lived in Greece. The most renowned Greek of those long-ago days was a blind man who had been a slave. His name was Homer and he was a poet and a storyteller. Homer's stories were so good that we still read them today. They are stories of real heroes and heroines and of mythological gods and goddesses. Homer's two adventure books, the *Iliad* and the *Odyssey,* are among the best books ever written.

The Greeks had an idea for governing people that hadn't been tried before. They called it democracy. In the Greek language, *democracy* means "people's rule." In Athens, which was a powerful city-state, the Greeks tried democracy. They let the people (except for women and slaves) vote and rule themselves. Since most men didn't have time to vote on everything, they elected leaders to decide some things for them. That made their form of government a *republic*. A republic is a place where people elect representatives who govern them according to law. The Athenian democratic republic worked wonderfully well. Athens had a golden age of art and music and poetry and science.

A Greek painted this vase 2,500 years ago with a scene from the *Iliad*: the Greeks and Trojans fight over the fallen body of the hero Achilles (in the middle).

More than 2,000 years later, when people in colonies on the shores of North America decided to form their own government, they read books by Greek writers and came up with a democratic republic called the United States of America.

The Romans were another Mediterranean people. A Roman writer named Virgil took Homer's stories and some other tales and put them in a book called the *Aeneid* (uh-NEE-id). (That book is very good reading.)

Roman generals, like Julius Caesar, conquered much of the Mediterranean world. Then they had to govern that world. Roman thinkers spent a lot of time talking and writing about ways to run a government. For a while they, too, lived under a republican government. The men who wrote our constitution knew all about the Roman republic, too.

A Greek writer named Plato wrote an important book about government called *The Republic.*

The Romans were ruling most of the Mediterranean lands when a Jewish boy, named Jesus, was born in Israel. Jesus lived in a town called Nazareth. From his preaching and his example a great new religion grew. That religion was Christianity. The ideas of Jesus are written in the New Testament. Jesus Christ was born about the year 1. Modern history dates from the birth of Christ.

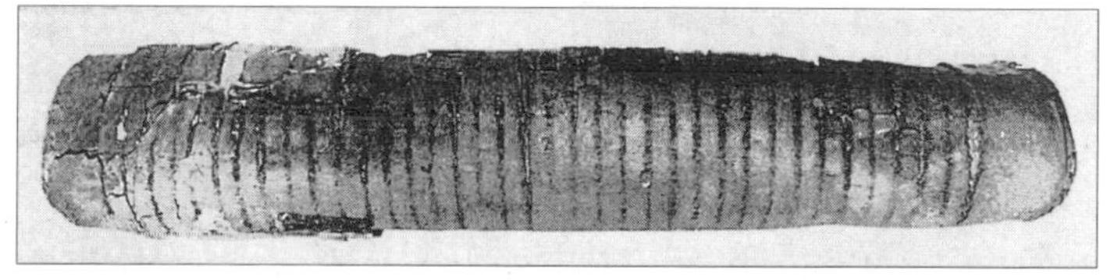

In 1947, an Arab herdsman was grazing his flock near the Dead Sea in Jordan. He went after a stray goat and stumbled into a cave, which contained some old, old jars and baskets. In them, wrapped in linen, were ancient scrolls, written in Hebrew, in Greek, and in Aramaic, another ancient language related to Hebrew and Arabic. Later more caves and scrolls were found. The Dead Sea Scrolls are the oldest known copies of parts of the Hebrew Bible. They were probably written before the birth of Christ, sometime between 150 B.C.E. and 40 C.E.

Another child was born in the land to the east of the Mediterranean. He was Mohammed, a descendant of Ishmael, and he was born in the city of Mecca in the year 570. Look at a map and see if you can find Mecca. Mohammed founded the religion known as Islam, and his teachings are found in a book called the Koran. People who practice Islam are known as Mohammedans or Muslims.

Islam spread rapidly across Arabian lands and into North Africa and the Sudan. In the year 711 an African general, Taril ibn Ziyad, sailed an army 13 miles from Morocco across the mouth of the Mediterranean Sea and landed on a Spanish hill. That rocky hill became known as Jabal Taril (Taril's hill), which was soon shortened to Gibraltar.

Taril's army of Muslim Africans fought and defeated an army led by King Rodrigo of Spain. For the next seven and a half centuries most of Spain was ruled by Muslims (the Spaniards called them Moors). It was a splendid era in Spain. It was a time of religious tolerance: Jews, Muslims, and Christians lived together in harmony. The Moors built great cities with centers of learning. Agriculture flourished in Spain as

In a feudal system, the peasants have to declare their loyalty to a lord and do a certain amount of work for him. In return, the lord is supposed to protect the workers and help them in bad times. Working for a lord like this was a bit better than being a slave, but not much.

it never had before or has since. We call these times the Middle Ages, because they lie between ancient days and modern times. During the Middle Ages, most Europeans were Roman Catholic Christians and religion was the center of their lives.

It was a time when knights, lords, and ladies lived in splendid feudal castles. And religious crusaders went adventuring in foreign lands. But most people weren't lords or knights, and most didn't go on crusades. Most people were peasants. They worked on the land, and their lives were short and hard. They couldn't read, they didn't get a chance to travel, and they didn't know much. Books were kept in monasteries and were read only by monks. Most Europeans had forgotten the arts and learning of the Greeks and Romans.

A ***monastery*** is the home of a group of monks. In the Middle Ages monasteries were big and powerful. They had their own farms, workshops, and schools, and usually didn't have to buy anything from outside.

It was different in Spain. Remember, Arabs, Jews, and Christians had brought their scholarship, arts, and energy to that land. Spain was thriving. Then the Moors started fighting among themselves. In the 15th century their armies were defeated by Christian armies; they were driven from Spain. The Jews, who had helped rule Spain, were forced to become Christian or flee. But the Moors had developed modern sailing technology, and that allowed Spain and Portugal to build the caravels used in the Great Age of Discovery.

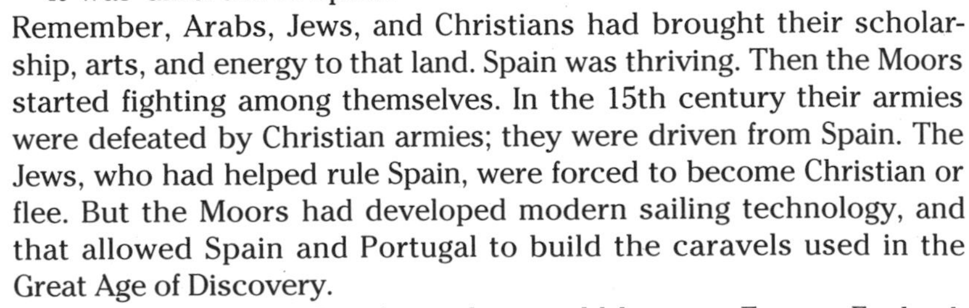

Caravels were ships first built in the 15th century. They made long voyages possible, because they could sail into the wind.

In other regions—the places that would become France, England, Germany, and Italy—new groups of people began taking power. Cities were growing and trade was expanding. The Middle Ages were being replaced by a Renaissance (REN-uh-sahnce)—a rebirth of learning and art. It was the forgotten ideas of the ancient Greeks and Romans that

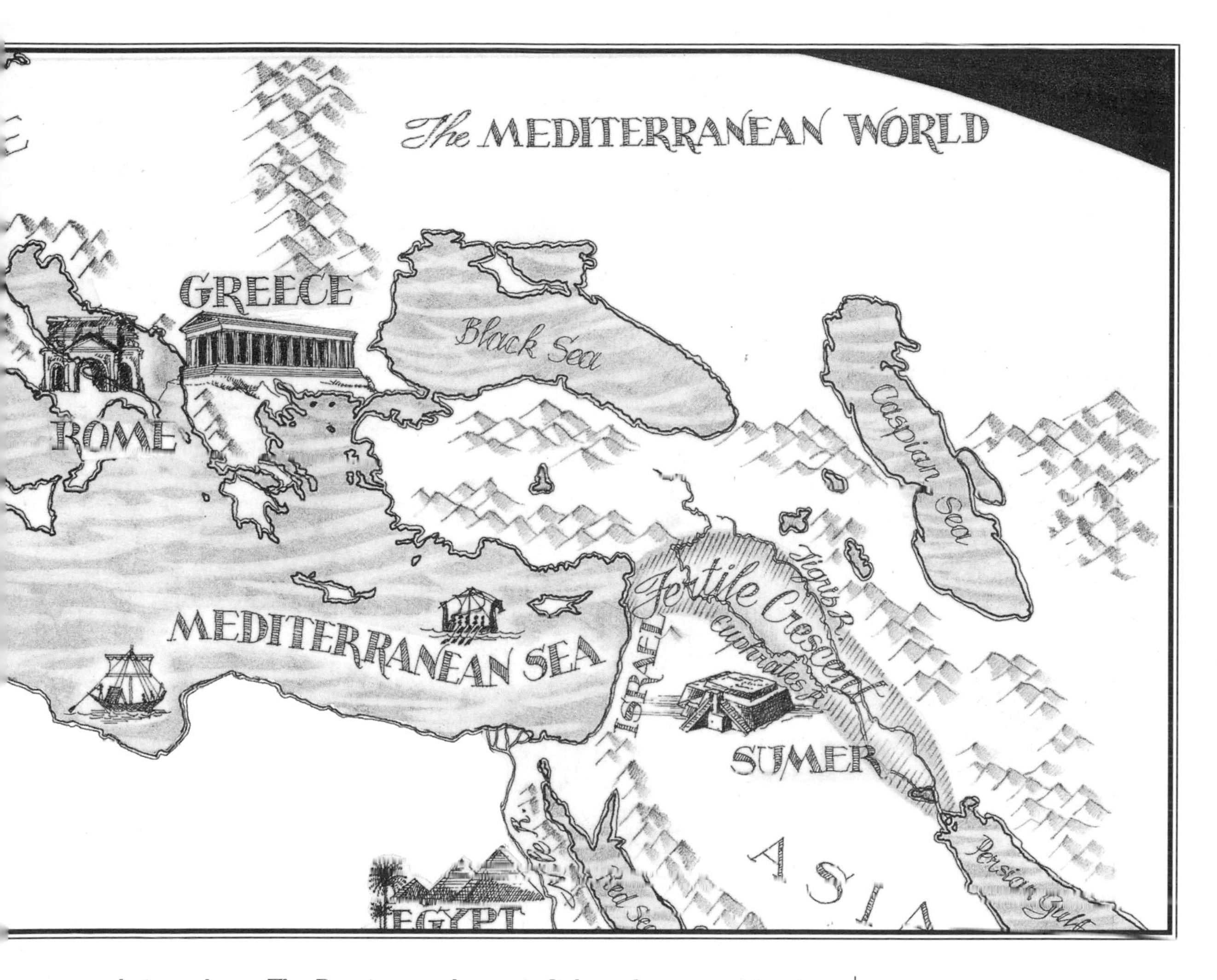

were being reborn. The Renaissance began in Italy and soon was inspiring thinkers all through Europe. At the same time, travelers were describing the fabulous civilizations in the Far East. Before people could even digest all the fresh ideas, two whole new continents—South and North America—were discovered. And seagoing explorers proved, beyond anyone's doubt, that the world was round. As if that weren't enough, the Catholic church was rocked by reformers (people who wanted to change the church), and that led to a host of new Christian churches, called Protestant. And the European nations (that we know today) were being formed. As you can see, it was an exciting time, but confusing, too.

The ideas of all these peoples—with all their religions and cultures—made up an idea pool, a kind of cultural stew, that sailed to America across the ocean from Europe. (Both the good ideas and the

Marco Polo, a trader from Venice, went to China in the years 1275–1292. He saw wonderful things, met the Grand Khan, and came back and wrote a book about it all.

This sign was the Totem of the Five Nations. In 1570, five Indian tribes, the Mohawk, Oneida, Cayuga, Onondaga, and Seneca, united in an Iroquois confederation.

bad came to America. One bad idea was slavery. Some of the people we have talked about kept slaves.)

In America there were already civilizations as ancient as those that had begun in Ur and Egypt and Greece. The Indian civilizations surprised the Europeans. To begin with, they weren't all alike. Some Native Americans were ruled by powerful lords. Some practiced slavery. But others lived democratically. The Iroquois had a government of the people. Women were important in Iroquois society, and there was much freedom.

There were no democracies in Europe. There were kings and emperors. The republics of Greece and Rome were a distant memory. But in America there were republics—Indian republics. Right away, some of the newcomers were impressed with the free life the Indians led. They thought about that free life and added it to their idea pool.

Unfortunately, many of the newcomers didn't understand the Native American cultures. Because they were different, the Europeans thought them inferior. They called the Indians savages. They destroyed much of the Indian heritage before they came to appreciate it.

The European newcomers learned more from the Native Americans than they realized. They learned ways of planting, harvesting, and hunting. They added Indian words to their languages. They exchanged foods, animals, diseases, and even weeds. Indian ways enriched the new culture that was forming. That culture became a mixture: Africans came—in chains—but that didn't stop them from adding ideas, energy, stories, music, and knowledge of agriculture. Then people came from Asia, bringing the wisdom of ancient religions and cultures.

Out of that worldwide mixture a nation developed—our nation, the United States—and a new civilization, an American civilization, where peoples of East and West and of many races and religions now live together and share ideas.

What does that mixing do for us? It makes life in the United States very interesting. People in America eat pizza, chow mein, and pita bread—sometimes in the same meal. In America, children whose ancestors left England in tiny wooden boats go to school with children who themselves left Vietnam in tiny wooden boats. In America, children who worship in a church go to school with children who worship in a mosque or temple or synagogue or kiva. In America, people who have their own beliefs—rather than those of an organized religion—are given respect.

Our nation began as an experiment. Nothing like it had ever been tried before. The citizens of most other countries couldn't even imagine living with people who seemed different from them. Can people who don't look alike build a free and equal society together? Read on, and see for yourself.

A Great Mystery

A Dakota Indian named Luther Standing Bear would think about the meeting of Native Americans and Europeans and write:

While the white people had much to teach us, we had much to teach them, and what a school could have been established upon that idea!...Only the white man saw nature as a "wilderness," and only to him was the land "infested" with "wild" animals and savage people. To us it was tame. Earth was bountiful and we were surrounded with the blessings of the Great Mystery.

1 A Sign in the Sky

Experts disagree about whether Halley's was the comet seen in 1066. Some say it was; others say it may not have been the same comet.

In 1607 a dazzling comet lit the sky over Europe. "A comet, a comet!" people cried and pointed to the heavens. In those days almost everyone could name the bright stars and planets. They knew the constellations and the mythical stories of their forming. Mothers and fathers showed their children the stars named for the dog Sirius, or Draco the Dragon, or Cygnus the Swan.

Pollution and city lights had not yet dimmed the skies. Nor was there much else in the evening to capture attention. Few could afford costly candles. So when something out of the ordinary appeared in the night sky, people saw it and wondered at its meaning.

These were times when most questions were answered by religious faith or superstition. Modern science was just being born. Stars were thought to be the lights of heaven, and comets were said to be messengers sent to foretell danger and dire change.

Many who watched the bold comet were frightened. They might have been even more fearful had they known this was the very comet that had been seen in 1066, when the Norman kings invaded England from France. Those French conquerors had changed England, and the English language, for all time.

But you didn't need a comet in 1607 to see that Europe was changing. The old religion—Catholic Christianity—had broken apart. England had become Protestant, then Catholic, then Protestant again. Now some people, called Puritans, were saying the country wasn't Protestant enough. It was all very disturbing to people used to a secure faith. To make matters worse, there were economic problems, too

In the great Catholic nation of Spain, the government was bankrupt. Although nobody knew it, Spain's glory days were over. Would that arrogant little Protestant island—that England—become Europe's new

This is a diagram from Copernicus's *On the Revolution of the Celestial Spheres.* It was the first time ever (1543) that the sun (*Sol*) was shown at the center of the solar system.

Even when Galileo was made to recant (deny) his belief that the earth moved around the sun, he didn't really deny it. Afterward, it's said, he muttered quietly, "But it still moves."

Opposite is an illustration of the path of Halley's comet. At the top is its orbit through the solar system; below is its path across the sky, in relation to the constellations (see Leo's paws?). Halley tracked his comet less than a century after Galileo was jailed for saying that the sun was the center of the solar system. Yet this diagram tells us Galileo's belief was now common knowledge, accepted by most educated people.

leader? Now that England's magnificent Queen Elizabeth was dead, no one knew where England was heading.

In this world that had once seemed orderly, ideas were changing. New thinkers, like the Italian scientist Galileo Galilei, were actually saying that Aristotle, the greatest of scientists, had some ideas that were wrong! Galileo even whispered that Nicholas Copernicus might be right. Copernicus, a Polish astronomer, had said the sun, not the earth, was the center of our universe. How could that be? Everyone knew that the planets and stars and sun all revolved around the earth. If that idea was wrong then the Pope and all of Europe's rulers were wrong.

Of course, that disturbing idea got Galileo into a lot of trouble. It ended up changing everything people believed. Change is troublesome, especially to those in power. And yet the new ideas, like germs, seemed to travel on invisible wings. The epidemic of thought was soon out of control.

In 1609, just two years after the comet appeared, Galileo built one of the world's first telescopes. (It was a lot stronger and therefore more useful than the few made before it.) Galileo took much of the mystery from the skies and replaced it with scientific order. When the comet of 1607 came again, in 1682, an English scientist named Edmond Halley tracked it and learned that it takes more than 75 years to complete its trips around the sun. Halley predicted that the comet would return in 1758, and he was right.

But in 1607 people knew none of that. They didn't know the year 1607 was to become famous. That very year the seeds of a new nation—a new way of governing, a new way of looking at the world—were being planted on the North American continent by a small group of English men and boys.

The message that some people read in the comet was right: the world was in for astonishing changes. Would they be changes for the better? Not for those people called Indians, who would soon meet the pale-skinned English. For them the changes were tragic.

But for England, and the rest of the world, 1607 marked the beginning of what would turn out to be an awesome, momentous, earth-shaking experiment. An experiment that would lead to democracy and to a government dedicated to liberty and the pursuit of happiness.

Let's watch the United States happen. Take yourself to the beginning of the 17th century, to England, where some brave men are getting ready to travel to a place they call the New World. They have no idea what is ahead of them. They'll bring along their Old World ideas—good and bad. Pack a bag, we're going to join them.

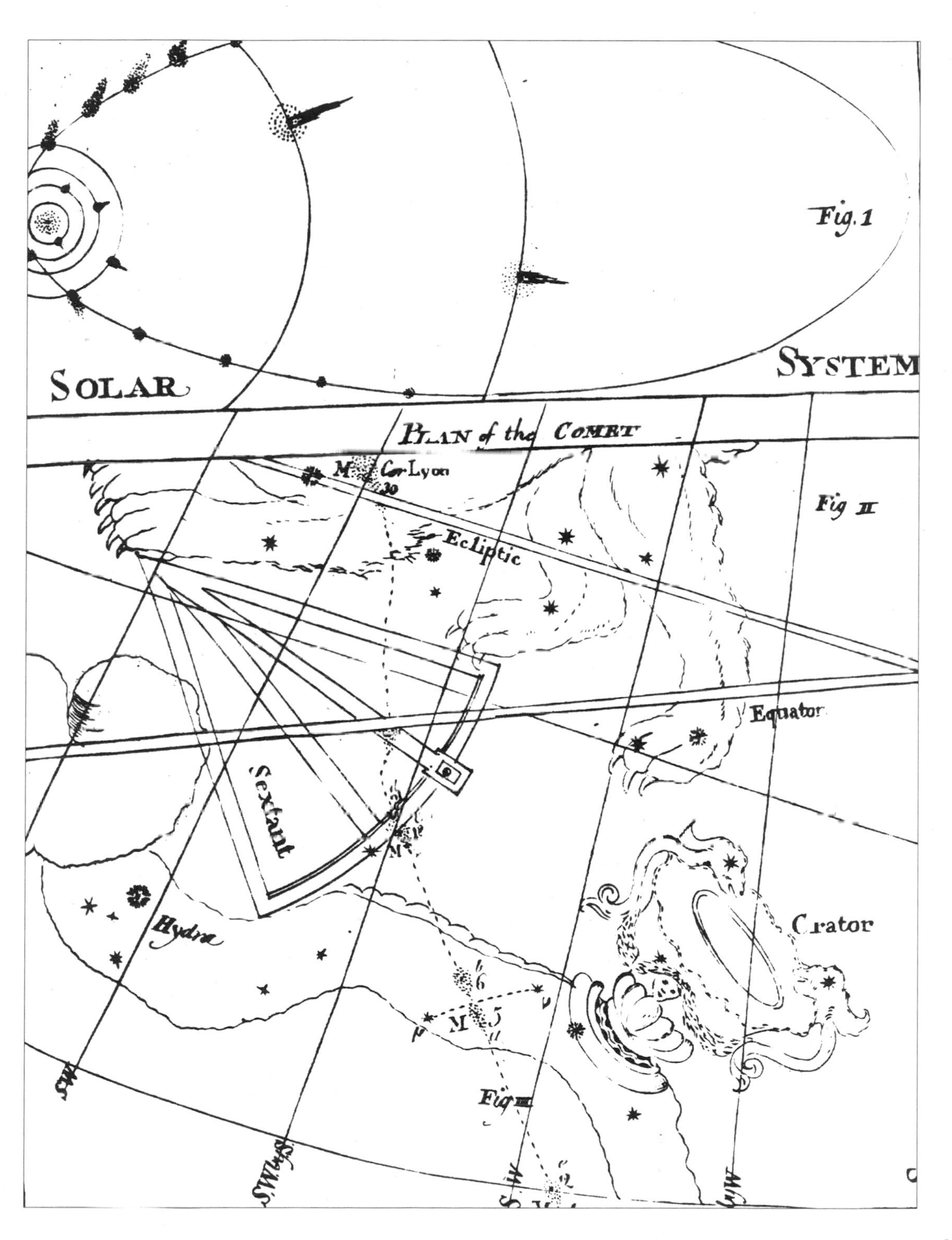
Fig. 1
SOLAR
SYSTEM
PLAN of the COMET
Cor Lyon
Ecliptic
Fig II
Equator
Sextant
Hydra
Crator
Fig III

2 Across the Ocean

English ships of the 17th century were sturdy but tiny. An ocean voyage in one seemed endless.

On the docks of the river Thames (TEMS), near London, a group of Englishmen readied themselves for a trip across the vast ocean. They were brave men, as they had to be to embark on an adventure such as they had in mind. They were going to a New World, where there were strange animals, deep forests, and people said to be wild.

To ***embark*** means to set off.

There was also gold. They were sure of that. A play, popular in London, told of Virginia. It said:

> *I tell thee, gold is more plentiful there than copper is with us.... Why, man, all their dripping pans and their chamber pots are pure gold....and as for rubies and diamonds, they go forth on holy days and gather them by the seashore, to hang on their children's coats and stick in their caps.*

Of course the authors of that play had never been to Virginia, but people believed them. Everyone knew there was gold in America. Hadn't the Spaniards found mountains of gold? And wasn't England now the greatest of nations? Her time had come.

About 144 men actually set sail for America, but the sailors weren't counted because they returned to England.

Of the 105 men who stood on the docks, more than half listed themselves as "gentlemen." In England gentlemen were not expected or trained to work. They lived on family money. They had time for adventure; they hoped to find riches.

Most brought their best clothes for the trip: their puffed knee pants, their silk stockings, their feathered hats, and their gaudy blouses.

Gaudy means showy and brightly colored.

Those in plainer clothes were the gentlemen's servants; a few were carpenters and bricklayers; four of them were boys—probably orphans or runaways. The boys were called *younkers* and were expected

In dirty, crowded London it was easy to imagine that America was an untouched Garden of Eden. It had to be full of animals known to Europeans only from myths—but which most people firmly believed existed somewhere.

A ***peril*** is a danger or risk. A ***privateer*** was a pirate with a government license. Privateers split their loot with the king.

to climb the rigging, high on the ship's mast, to help set the sails and keep a lookout for land and danger. If a younker fell into the ocean and was lost—well, too bad. That was one of the perils of sea travel.

Ocean travel was risky—they all knew that. They also knew that their captain, Christopher Newport, was one of England's finest sailors. As a privateer, he had sailed the New World's seas. Queen Elizabeth had encouraged English captains to prey on Spanish ships. And Newport had led an expedition that destroyed or captured 20 Spanish vessels and sacked four towns in the West Indies and Florida. He was an English hero. What the Spaniards thought of him is something else.

But the men and boys who climbed onto three small ships and set sail down the river Thames felt confident with Newport in command. What surprised them all were contrary winds.

They sailed into the Atlantic, and the winds blew them back to England. Out they went again—and back. For six weeks those strange winds blew, while the voyagers ate their precious food and got nowhere. Now there were grumblings. Some wished they'd never come. Do you think they were scared? Do you think they thought of turning back? Pretend you are a younker. Are you excited? Or afraid? Or both?

Short Rations

The ships of the day were brightly painted with striped masts, banners, and flags. But there was never really enough food aboard for the six- or eight-week voyage. Those who made it across were usually weak and hungry when they landed—if not dead or sick with scurvy. A sailor's song went like this:

We ate the mice, we ate the rats,
And through the hold we ran like cats.

This is supposed to be a three-toed sloth. The artist had probably never even been to America, let alone seen a sloth, so he made up an idea of it from someone else's description. He must have thought it looked like a bear with a man's head. This was the kind of beast the early colonists expected to find.

Abundance is plenty.

They had boarded ship in December 1606, but it was February of 1607 when, finally, they lost sight of England. Captain Newport soon had them in the Canary Islands, where the three ships took on fresh water and food. Then they were off to the islands of the West Indies, where they rested and prepared themselves.

For they understood, when they left the West Indian island of Martinique, that they were heading to little-known territory. They were to do something Englishmen had not done before: they were to start a colony on the mainland. Spain had grown rich because of her colonies. England would beat Spain at that game. These Englishmen were determined to enrich England—and themselves, too.

For a few of the voyagers there may have been something else besides gold that drew them to America: that was its beauty and abundance. Those who had seen the land wrote of birds and flowers and fish more gorgeous than one could imagine. A poet called Virginia "earth's only Paradise."

England seemed crowded. Timber was scarce and getting scarcer. Farmland was disappearing. London's streets were filled with beggars.

Might America's land and trees and soil be as valuable as the gold nuggets the adventurers thought they were sure to find? Could this New World become a land of hope and opportunity? There were some in England who thought so.

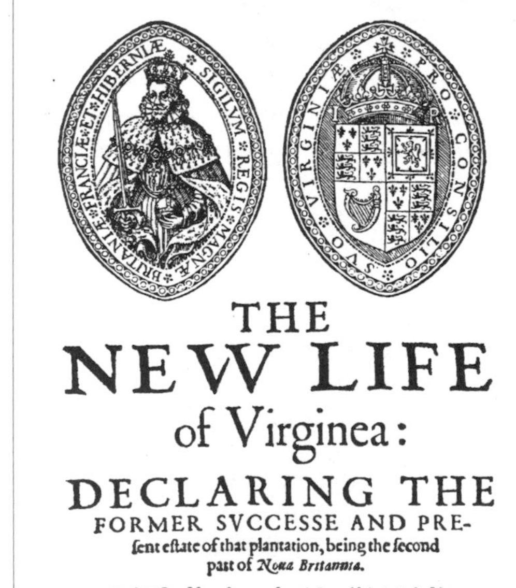

THE
NEW LIFE
of Virginea:
DECLARING THE
FORMER SVCCESSE AND PRE-
ſent eſtate of that plantation, being the ſecond
past of *Noua Britannia.*

Publiſhed by the authoritie of his Maieſties
Counſell of *Virginea.*

LONDON,
Imprinted by *Felix Kyngſton* for *William Welby*, dwelling at the
ſigne of the Swan in Pauls Churchyard. 1612.

The colonists were certainly headed for a new life in Virginia—but not always a good one. Fliers like these were put out to attract more settlers. Sometimes they weren't very truthful.

3 The First Virginians

The Powhatan had about 9,000 subjects; his land stretched from what is now Washington, D.C. to northern North Carolina.

When English parents told stories to their children, they often spoke of monsters, trolls, wild beasts, and witches. Those were savage stories, strange and disturbing. Since everyone knew there was a savage side to life, the stories had a kind of realness to them, even when they were make-believe.

But those weren't the only stories they told. There were tales of splendor and goodness, too. Every child heard the Bible's first story, which is of a Garden of Eden. Eden was a place of great beauty, a paradise. Many of the goodness stories were about sweet, simple people who lived in harmony with nature. And those stories seemed real, too, because there was much goodness in the world.

Did you know that the stories you read in childhood stay with you all your life? They influence adults more than most of them realize. So when Englishmen and women learned of a land of great beauty, where people lived close to nature, many of the English thought of that land as paradise. They called the natives savages, but meant the word kindly. The first English visitors to the New World described the Indians as "courteous" and "gentle" and "great."

But, later, when others met those great savages, they found they didn't always act as people do in storybooks. Soon some were calling them worse names than "savages." They called them "beasts." Some said they were servants of the devil. Others said they were part animal and part human.

But those people—the Indians—were just real people, like the English. They lived in families, in towns, governed by leaders. They farmed, hunted, played games, and fashioned beautiful objects. Some

Earthly Paradise

We have discovered... the goodliest soil under the cope of heaven, so abounding with sweet trees, that bring such sundry rich and pleasant gums, grapes of such greatness, as France, Spain, nor Italy have no greater....the continent is of huge and unknown greatness, and very well peopled and towned, though savagely, and the climate so wholesome, that we had not one sick since we landed here.

It was letters like this one, written by Ralph Lane in 1585, that made people want to go to the New World. Lane was a member of Raleigh's Roanoke Colony.

An ***estuary*** (ESS-tew-air-ee) is the body of water where the mouth of a river meets the sea.

The bear grease that the Indians rubbed on their bodies made their brown skin shine with a reddish glow. Europeans thought it really was red, and that was how the name "redskin," which many Europeans once used for Native Americans, came to be. It's based on a *misperception*, which means an error in the way things are seen.

Remember, horses first came to America with the Spanish conquistadors. Even 100 years later they were still a rarity.

of them were wise and some were foolish. Some were kind and some were mean. But most were a bit of all those things.

One of the most interesting Indians the English would meet was the Powhatan, the ruler or emperor of eastern Virginia. His real name was Wahunsonacock, and he had inherited an empire of five tribes. Through daring, strength and leadership, Powhatan soon held sway over dozens of villages and thousands of Indians. The English would call the Indians of his empire Powhatan Indians.

They were Woodland Indians, who spoke Algonquian (al-GON-kwee-un) dialects and hunted, fished, and farmed in a region of great abundance. The area surrounded the Chesapeake Bay and went west to mountain foothills and south to what would someday be North Carolina's border. It was a land of rivers, bays, and estuaries; of ducks, geese, wild turkeys, and deer; of fertile soil, fish, and shellfish; of wild berries, nuts, and grapes.

Powhatan's people raised vegetables—corn, beans, squash, and pumpkins—which was more than half the food they ate. Because they farmed, they lived in settled villages. Corn was their most important food. They ground it and made it into flat pancakes that served as bread or rice does in many other cultures. Aside from corn, the food they ate changed with the seasons: fresh vegetables in summer and fall; game in winter; and fish, stored nuts, and berries in the spring. (Spring was when corn supplies ran low and they sometimes went hungry.)

There was much small game in the region: raccoon, opossum, squirrel, turkey, and rabbit. But it was deer these Indians relied on most for food and clothing. Unfortunately, like people elsewhere, they overhunted; deer became scarce. And they knew if they roamed outside the Powhatan's territory—looking for better hunting grounds—they risked war with other tribes.

It was the men who hunted, fished, and fought. Women farmed. Men and women had set roles in this society and rarely changed them. Children helped their parents, played, and didn't go to work until they were young adults.

The boys often played in scarecrow houses that stood in the middle of the fields. From there they threw stones at rabbits or other animals that might nibble on the crops. It trained their throwing arms, and that helped when they became hunters. Little girls played with clay, made pots, and helped their mothers plant and cook. Boys and girls played running games. There were no horses (they hadn't arrived in this part of the New World yet), so fast runners were prized. Sometimes they would dress up like their parents—painting their bodies and wearing necklaces and bracelets of shells and beads and animal bones.

Men and women tattooed beautiful designs all over their bodies.

Men sometimes hung animal claws, birds' wings, bats, even live green snakes around their necks. They rubbed themselves with bear grease—it repelled mosquitoes, kept them warm, and made their skin glisten in the sunshine. Most of the year these Indians needed little clothing, although in winter they wore deerskin garments and, sometimes, cloaks of feathers or fur.

This is John Smith's map of Virginia, made in 1612. Can you find Jamestown? And Powhatan? You'll hear more about them—and John Smith—in the next chapter.

Werowance is a Delaware Indian word that means, literally, "he is rich."

Powhatan was said to have 100 wives. One third of his 9,000 subjects were warriors. Powhatan gave his decorated deerskin to Captain Christopher Newport. It is now in a museum in Oxford, England.

The Chesapeakes lived on land that is now included in the cities of Virginia Beach, Portsmouth, Chesapeake, and Norfolk.

The great Powhatan had a beautiful deerskin sewed with designs in lustrous pearls. Powhatan had stacks of deerskins and a storehouse of corn, and he had copper and pearls. The tribes brought all this and more to him. It was tribute given to a ruler.

Each tribe had its own leader, called a *werowance* (WEER-ah-wunts), and also priests and healers and others with power. But the Powhatan was special. All those who met him noted it. He knew how to command, when to be stern and unforgiving and when to be understanding. The story was told of the time he visited the Potomacs (puh-TOW-mucks), who were under his rule. The young Potomac warriors came before him, and each told of awesome deeds of valor against fierce enemies or wild beasts of the forest. Finally one young man stood before Powhatan and said, "I, my lord, went this morning into the woods and valiantly killed six muskrats. While that may be no more than boys do, it is true, while much you have heard is fable." When Powhatan heard that, he broke into laughter and gave a reward to the truth teller.

Powhatan's priests had foretold that his mighty empire would one day be destroyed by men from the east. The Chesapeakes, who lived by the ocean, were the easternmost of his tribes. Perhaps that is why, just before the 16th century turned into the 17th (in the European method of reckoning), Powhatan fought the mighty Chesapeake Indians and left them weak and powerless.

Powhatan didn't yet know that three small ships were heading for his realm. They were coming across the ocean from an island far to the east.

Virginian Indians cooking a stew of corn, fish, and beans. This picture was drawn at the end of the 16th century by John White.

4 English Settlers Come to Stay

The crown and coat of arms on the Virginia Company's seal show that it was granted by the king.

In Virginia, April is a sweet month. Strawberries and white dogwoods blossom below the green of tall pines. Redbuds are emerging, and so are grape leaves, honeysuckle, and wild roses.

It was thus in April of 1607, when three small ships landed at the mouth of the Chesapeake Bay. The ships were the *Susan Constant,* the *Discovery,* and the *Godspeed*, and they had been sent from England by a business corporation called the London Company.

They were the same three ships that left the docks on the river Thames. The voyagers had been told to look for gold and a river or passage that would go through the country to China and Japan. They were also to see if there were other ways to make money on this unknown continent.

The ships anchored near an elbow of beach they named Cape Henry (in honor of young Henry, the king's oldest son). Some of the mariners rowed to shore and set out exploring. "We could find nothing worth the speaking of, but fair meadows and goodly tall trees, with such fresh waters running through the woods, as I was almost ravished at the first sight thereof," wrote George Percy, who was one of the gentlemen adventurers.

On their way back to the ship the Englishmen were attacked by Indians, who came "creeping upon all fours...like bears, with their bows in their mouths," but when "they felt the sharpness of our shot, they retired into the woods with a great noise."

The local Indians knew about white men, and they didn't want them around. Spain—England's old enemy—had tried to start two colonies in the Chesapeake Bay area. An Indian prince from the region

From the ***instructions of the London Company to the First Settlers:***

"When it shall please God to Send you on the Coast of Virginia you shall Do your best endeavour to find out a safe fort in the Entrance of some navigable River making Choice of Such a one as runneth furthest into the Land, and if you happen to Discover Divers portable [navigable] Rivers and amongst them any one that hath two main branches if the Difference be not Great make Choice of that which bendeth most towards the Northwest for that way shall You soonest find the Other Sea."

John Smith's portrait labels him "Admirall of New England." Even if it wasn't an official title bestowed by the king, Smith deserved it. In addition to exploring and mapping much of the territory of Virginia, Smith sailed to New England, charting its coast from Maine to Cape Cod.

had been taken to Spain, baptized a Christian, educated, and returned to his people. That prince had far more schooling than most of the Englishmen who now wished to invade his land. The Englishmen didn't seem to know any of that. They thought of the Indians as savages.

The Englishmen spent a few weeks exploring the bay area. They feasted on strawberries ("four times bigger and better than ours in England"), ate oysters ("which were very large and delicate in taste"), and noticed grapevines ("in bigness as a man's thigh"). The oysters and mussels "lay on the ground as thick as stones; we opened some,

Today "admiral" is spelled with one l. *Do you see any other old-fashioned spellings in this portrait?*

and found in many of them pearls...as for sturgeon [there were so many of these fish] all the world cannot be compared to it."

They planted a cross at Cape Henry, thanked God for their safe voyage, and watched as Captain Newport opened a sealed metal box. The box had been entrusted to him by the London Company. (Newport would soon sail back to England.) He opened the box and read six names and his own. They were to be members of a council and elect a president. One of the names was a surprise. It was John Smith; he was locked up in the ship's belly.

Smith was of yeoman (YO-mun) stock—and feisty; he was not one of the gentlemen. He had angered some of those gentlemen and they had put him in chains. They were planning to send him back to England. Now they would have to work with him.

A ***yeoman*** was a small farmer who cultivated his own land (it was the area of land that was small, not necessarily the farmer).

The instructions in the box said they were to go inland, up a river, and find a suitable place for their colony. So they left the mouth of the Chesapeake Bay and sailed up a river they called the James, to a site they named Jamestown.

Several men, including John Smith and Christopher Newport, went on, up the James River, in search of a passage to China. They had no idea of the size of the country. When they saw breaking waves, they were sure they had found the western coast and the Pacific Ocean. John Smith wrote in his log of the "ocean ahead."

They soon discovered that the waves were caused by water tumbling over rapids in the river. They were at a site that would someday be the city of Richmond. The river would not let them go farther.

All those Jameses—the James River and Jamestown—were named to honor the new king, James I. When Queen Elizabeth died in 1603, still unmarried and childless, her great-nephew, James, was brought from Scotland to become king of the United Kingdom of Scotland and England.

James I wasn't a bad man. He was very learned about many subjects, but couldn't deal with people very well. He was kind and generous to his friends, but tactless with those he didn't know. He disapproved of tobacco smoking because he thought it filthy and disgusting, but he never washed himself—and he had terrible table manners.

King James had worked out a kind of deal with the Spaniards. It went like this: we English will stop raiding your ships, if you Spaniards will promise not to attack our settlers. So the new settlers weren't as worried about Spanish attackers as they might have been in the 16th century. As it turned out, there may have been Spanish spies among them.

But it was gold that was on their mind when they reached Jamestown, and they soon began searching for it. They also built rough huts for shelter and a triangular fort for protection.

Jamestown was almost an island, with a narrow sandbar link to the mainland. It would be easy to defend against Indian raids or against ships, just in case the Spaniards did decide to come up the river. Besides, deep water touched the land. They could sail right up to the site and tie their ships to trees.

Christopher Newport sailed back to England. The adventurers were on their own in America. Newport thought he had gold when he took a barrel of shiny earth back to England. It turned out to be "fool's gold" (iron pyrites).

As it turned out, they couldn't have picked a worse spot. The land was swampy, the drinking water was bad; it was hot in summer and bone-chilling in winter. The mosquitoes drove the settlers crazy and carried malaria germs.

They might have handled all that if they had been a decent bunch. But, for the most part, they were lazy and vain and fought among themselves. And their first two leaders were incompetent—which means they made a mess of the job.

All were men; they brought no women. Remember, most were gentlemen, with no training or taste for hard work. To be fair, they had been misled about the New World. They expected to find gold at their feet, and they wasted valuable time looking for it. And there wasn't a farmer among them.

To make things worse, the London Company, which had paid for the voyage, showed poor sense. It gave all the colonists salaries and did not allow them to own property. No one had a reason to work hard, because the hard workers got the same pay as those who did nothing.

Besides all that, they had bad luck—lots of bad luck. The worst may have been that they brought some English germs across the sea. One was a typhoid fever germ that killed many of them. Tidewater Virginia had other germs (especially dysentery germs) that made some sicken and die. The Indians killed still others. Some starved. What happened to those eager men and boys who had stood on London's docks in December? Fewer than half of them saw another December.

Yet the news wasn't all bad. This was the first English colony that survived in the New World. A few things had to go right to make that happen. One man, more than any other, helped make things go right. He was short, scrappy, red-bearded John Smith—who had come to Virginia in chains. He was Jamestown's third president and a born leader, even though many of the voyagers didn't like him.

The settlers built small houses of wattle and daub—sticks and clay. They finished just in time to stave off their first Indian attack on May 26, 1607.

From *The Proceedings of the English Colony in Virginia*, 1612: "There was no talk, no hope, no work, but dig gold, wash gold, refine gold, load gold, such a bruit [noise] of gold, as one mad fellow desired to be buried in the sands, lest they should by their art make gold of his bones."

5 John Smith

John Smith arranged the exchange of an English boy, Tom Savage, for an Indian boy named Namontack. Savage became an adopted son of Powhatan, learned the Algonquian language, and later was able to translate for the colonists. Namontack went to England with Christopher Newport.

Smith's Turks' head coat of arms. The Latin motto means "Vanquish and live."

The Jamestown Colony might not have survived without Captain John Smith. He was a tough, no-nonsense man who worked hard and expected everyone else to do the same. Some people admired him; others hated him.

Pocahontas admired him and saved his life—twice. She was a bright-eyed Indian princess who was about 12 years old when the settlers first came to Virginia. The pale-skinned men and their strange ways intrigued her and she came to visit them often. Sometimes she just turned cartwheels in the middle of the Jamestown settlement; sometimes she brought food. It was John Smith who seemed to interest her most. In him she recognized a person whose intelligence and curiosity matched her own. Like her father, the great Powhatan, Smith seemed fearless.

Many of the settlers hated Smith, but they recognized that he was a leader. When he went back to England, even his enemies missed him. They said he was a braggart and that he couldn't possibly have done all the things he said he had done. He might have exaggerated a bit, but he really did do those things—like selling his schoolbooks and running away to sea. Or going off to Hungary to fight the Turks. There, on one bloody afternoon, Smith beheaded three Turks. A grateful Hungarian prince granted him a coat of arms with three Turks' heads emblazoned on it.

He wasn't a Hungarian hero for long; he was captured and sent as a slave to Constantinople, where a Turkish woman bought him. But her relatives didn't think much of him, and he was sold again. This time he killed his master, escaped, got thrown into the Mediterranean Sea, wandered through Russia, Poland, and Germany doing heroic things, and ended up in North Africa fighting pirates.

That first summer in Jamestown the settlers barely made it. "Our drink was water, our lodging, castles in the air," they reported. "With this lodging and diet, our extreme toil in bearing and planting palisades strained and bruised us....From May to September those that escaped lived upon sturgeon and sea crabs. Fifty in this time were buried."

Those early days weren't all misery. After their first Christmas in Virginia, Captain Smith wrote, "We were never more merry, nor fed more plenty of good oysters, fish, flesh, wild fowl and good bread, nor never had better fires in England than in the dry smoky houses of Kecoughtan."

Naturally someone who liked adventure the way John Smith did would be attracted to the adventure of a new world. Besides, like Sir Walter Raleigh, he had an idea that America could someday be an English land. He understood that there was more to be gained in the New World than gold. He realized that there were great opportunities for men and women with energy and courage.

John Smith was 28 when he took over the leadership of the Jamestown colony, and things were in a bad way. This was his motto: "If any would not work, neither should he eat." There were some grumblers, but everyone wanted to eat. So everyone worked.

Smith went and got food from the Indians. He learned their language and he learned the ways they hunted and fished. He had been a soldier and he was tough, but he was also fair and honest; the Powhatans soon understood that. They respected him and he respected them. They called him *werowance*, or chief, of Jamestown. And that was what he was.

The Native Americans seemed undecided about how to act toward these strangers on their land. What would you have done if you were a Powhatan? How would you have treated the English leader, John Smith?

He had goods they wanted—axes and shovels and blankets—so they traded with him. He was a natural trader. He took his English goods to their villages and he came back to Jamestown with boatloads of corn. They told him of the prediction: that men from the east would destroy their villages. He told them he came in peace.

Still, the Indians couldn't seem to make up their minds.

Sometimes they entertained Smith with dances and feasting. Other times they tried to kill him. Once he was brought before an Indian werowance and he expected to die. He pulled out his compass, showed how it worked, talked about the heavens and the earth, and soon had a tribe of friends. Another time he was taken to the great Powhatan, who seemed to have several Indian warriors ready to beat his brains out. But Pocahontas, who was the Powhatan's favorite daughter, came to his rescue. She put her head on Smith's, and Powhatan let him live. Was it a prearranged ceremony, or did Pocahontas actually save his life? No one knows for sure, but the Indians ended up adopting Smith

In a page from John Smith's *Generall Historie of Virginia*, Smith lies captive before the Powhatan while Pocahontas pleads for his life to be spared.

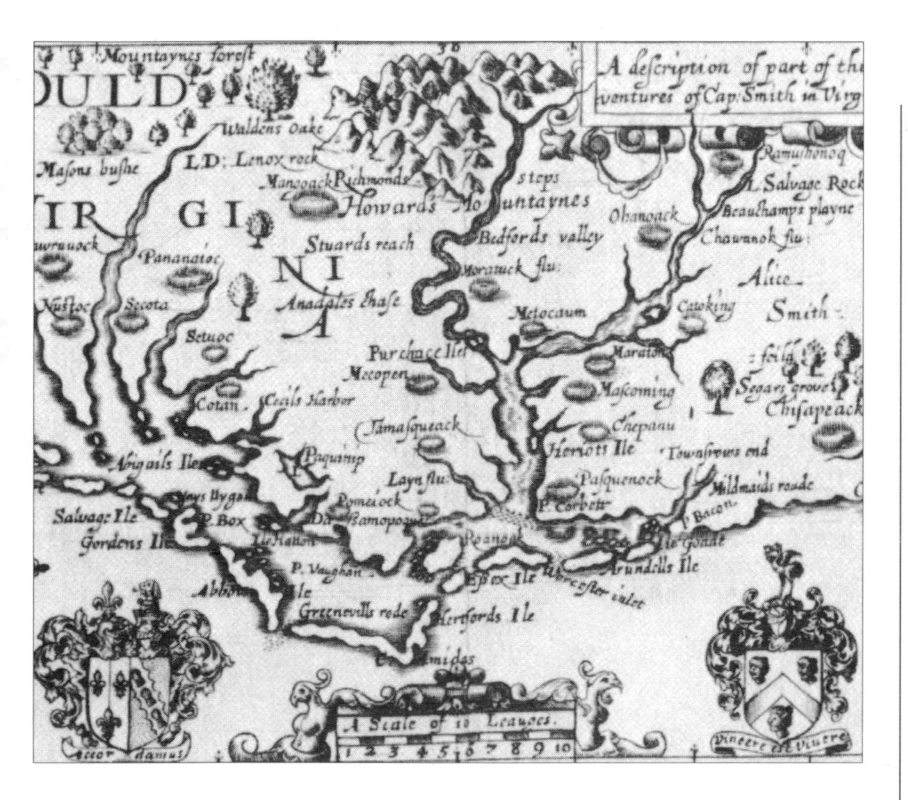

John Smith's map of "Old Virginia"—now the Carolinas. At the mouth of the big river in the middle is Roanoke Island.

into their tribe and making him an honorary chief. Now he was a member of Pocahontas's family.

When some other Indians tried to ambush Smith, Pocahontas warned him of the trap. Cats are supposed to have nine lives. John Smith had even more. While exploring the Chesapeake Bay, he was stung by a deadly stingray and was in such agony that he had his grave dug. He recovered—and ate the stingray. (The place where this happened, near the mouth of the Rappahannock River in Virginia, is now called Stingray Point.)

John Smith was asleep in his boat when some gunpowder exploded. It "tore the flesh from his body and thighs nine to ten inches square.... To quench the tormenting fire, frying him in his clothes, he leaped over board in the deep river, where...he was near drowned." Smith was so badly wounded that he had to return to England. Pocahontas was told he was dead.

John Smith never got back to Virginia. But he did get to New England, which he named, and he mapped its coast as he had mapped much of Virginia. He wrote many books and became famous because of them. "I am no compiler by hearsay, but have been a real actor," he said. And so he was. We still read his books today, because they are so interesting.

Rough Justice

In some parts of our country, when a criminal is found guilty of a serious crime, such as murder, he can be sentenced to death by the judge. A death sentence is called capital punishment. How did capital punishment start in this country? The English brought the tradition with them. (Some Indian tribes practiced capital punishment, too.)

It didn't take long for the settlers to use capital punishment—a few months. The first execution was in the Jamestown colony in 1607. The victim was not a common criminal, but a member of a prominent English family and one of the seven-man ruling council of the Jamestown colony. His name was Captain George Kendall. His crime was treason. The colonists believed Kendall was a spy for the Spanish. Kendall attempted to escape aboard the ship *Discovery*, but was captured and condemned to die. He was executed by a firing squad.

6 The Starving Time

It didn't matter how many pigs there were if you couldn't get out to kill them.

Before John Smith left for England, he counted the food in the storehouses at Jamestown. "Ten weeks' provisions in the stores," he wrote. It was October of 1609. Was there enough food to get through the winter? Smith seemed to think so. He expected the Indians to supply corn, as they had before. Besides, the settlers had hens, chickens, and goats—and so many pigs that a nearby island was called Hog Island. In addition to all that, the woods abounded with deer, rabbit, and squirrel; the river was thick with fish, frogs, and oysters.

There were new people in Jamestown, brought from England by Captain Newport. Two were women: Mrs. Thomas Forrest, the wife of a settler, and her maid, Anne Burras. Anne Burras's arrival led to a happy event, which a poet, Stephen Vincent Benét, imagined three and a half centuries later:

—And the first white wedding held on Virginia ground
Will marry no courtly dame to a cavalier
But Anne Burras, lady's maid, to John Laydon, laborer,
After some six weeks' courtship—a Fall wedding
When the leaves were turning and the wild air sweet,
And we know no more than that but it sticks in the mind,
For they were serving-maid and laboring man
And yet, while they lived (and they had not long to live),
They were half of the first families in Virginia.

The Laydons soon had a baby. Can you guess what they named her?

Still more colonists arrived. Now there were many mouths to feed, but most people were optimistic. Everyone thought Jamestown had seen the worst of its troubles.

Everyone was wrong. That winter of 1609–1610 was as awful a time as any in American history. It was called the Starving Time.

Captain George Percy, who was now governor of the Jamestown colony, said the settlers felt the "sharp prick of hunger which no man can truly describe but he who hath tasted the bitterness thereof."

They ate "dogs, cats, rats and mice," said Percy, as well as "serpents and snakes" and even boots and shoes.

> *There were never Englishmen left in a foreign country in such misery...Our food was but a small can of barley, sod in water, to five men a day...our men night and day groaning in every corner of the fort most pitiful to hear...some departing out of the world, many times three or four in a night; in the morning their bodies trailed out of their cabins like dogs to be buried.*

What happened? Some historians say the Starving Time was an Indian war against the English invaders. The Powhatan may have decided to get rid of the settlers by starving them. He wouldn't trade with them. He laid siege to Jamestown. That means armed Indians wouldn't let anyone in or out. The settlers couldn't hunt or fish. They could hardly get to their chickens and pigs. The gentlemen ate the animals that were inside the stockade—without much sharing. That made the others very angry. Soon there was nothing for anyone to eat.

Trapped inside the stockade by Indians, the settlers were so famished that some ate an Indian they killed. One man ate his dead wife.

A few escaped. "Many of our men this Starving Time did run away unto the savages, whom we never heard of after," Percy wrote.

In London the Spanish ambassador learned of the misery in Virginia. (Some said there was a Spanish spy at Jamestown, but a spy wasn't needed; the disaster was common news in London.) The ambassador urged the Spanish king to send a ship and finish off the English colony. The Spaniards could have done it easily. So could the Indians, who never went that far. (What do you think American history might have been like if either of those things had happened?)

Finally, in May 1610, two English ships tied up at Jamestown's docks. Of the 500 people who were in Jamestown in October, when John Smith left for London, only 60 were still alive.

The Laydons named their baby *Virginia*.

Would Powhatan have behaved differently if John Smith had been around? This is one of those historical questions that are interesting to think about.

When the new governor, Sir Thomas Gates, arrived in Jamestown after the terrible winter, it looked to him "rather as the ruins of some ancient fortification than that any people living might now inhabit it."

7 A Lord, A Hurricane, A Wedding

A ***flotilla*** is a group or small fleet of ships.
Unholsome is the old spelling. Today we would write it as ***unwholesome***. (See opposite page.)

On board ship the expression ***all hands*** means "all people."

Shakespeare wrote a play, *The Tempest,* after reading about the storm that wrecked the *Sea Venture* off Bermuda.

The people who walked off the two English ships in May of 1610 were horrified by what they saw. Jamestown was a wreck. Fear of Indians had kept everyone inside the fort through the cold winter. The settlers had been forced to burn their buildings to keep warm. By spring there were hardly any buildings left. The few survivors looked like skeletons. "We are starving!" they gasped.

Those on the ships had been through an ordeal themselves. On their way to Jamestown yellow fever broke out. There was no treatment for the disease then, so people died and their bodies had to be thrown overboard. They were barely over the epidemic when they ran into a fierce Atlantic hurricane.

Do you know what a hurricane at sea is like? Here it is: phenomenal, roaring winds; towering, crashing waves; fierce, crackling lightning; ear-splitting thunder. One ship, the *Catch*, went down; all hands lost. The *Sea Venture*, with sturdy Christopher Newport as captain, was wrecked on coral rocks. Luckily the rocks were within wading distance of the island of Bermuda. If you have to be shipwrecked, Bermuda is not a bad place to be. The voyagers set to work and built two new ships, the *Patience* and the *Deliverance.* They were eager to get to Jamestown and start a new life there. They imagined that the colony was thriving.

You know what they found at Jamestown. Are you surprised that they decided to leave? They had had enough of this New World. On June 7, 1610, everyone marched out of the wretched settlement, climbed aboard ship, said, "Goodbye, Jamestown," and headed for England.

They didn't get far. Lord de la Warr (what state is named for him?) was on his way up the James River with a big flotilla and 300 settlers. He was the new governor, and a good one. He made them turn around and start Jamestown again.

Lord de la Warr called Jamestown "a very...unholsome place." Still, he stayed. The colonists set to work. They cleaned, fixed, and built. The Indians continued to make their lives miserable, but now the settlers fought back. Then a few Indians and a few colonists began to trade. Some Englishmen began going up the rivers to trade with distant tribes. Tom Savage, who had lived as an adopted son of the Powhatan, was able to interpret for both peoples.

More ships came. Artisans and laborers arrived, together with "gentlemen of quality" and more livestock.

And still many died, not from starvation but from the diseases that abounded in the damp atmosphere at Jamestown.

Lord de la Warr was one of those who got sick. He went back to England, and Sir Thomas Dale took command. Dale understood the need for a healthier settlement. So one was built at a great bend of the James River, near the falls that John Smith had mistaken for the Pacific Ocean. The new settlement was named Henrico, for Henry, the king's oldest son. It had "two fair rows of houses," three "store houses," a hospital, and a "fair and handsome church." All of it was constructed in four months—Thomas Dale saw to that. Dale was a stern man, strict and religious. Anyone who swore, broke a rule, or didn't work got whipped. Three offenses and you were executed. You can see why everyone worked.

Henrico soon had a visitor. Unfortunately she was dragged there. Her name was Pocahontas.

The Indian princess was visiting some Potomac Indians when an Englishman, on a trading expedition, lured her onto his ship and wouldn't let her off. He took Pocahontas as a hostage to Henrico, held

Pocahontas had two names. Her real name was Matoax. It means "little snow feather." Pocahontas, her nickname, means "playful." Then Pocahontas married John Rolfe. She was baptized and given a third name: Rebecca.

What is a hostage? *Where do we hear the word today?*

Governor Dale's law code for Jamestown was tough.This was one rule: "No man shall dare to kill, or destroy any Bull, Cow, Calfe, Mare, Horse, Colt, Goate, Swine, Cocke, Henne, Chicken, Dogge, Turkie, or any tame Cattel or Poultry...without leave from the Generall, upon paine of death in the Principall, and in the accessary, burning in the Hand, and losse of his eares."

her there, taught her the Christian religion, and gave her a new name, Rebecca.

Now it happened that a young Englishman named John Rolfe had a plantation nearby. He fell in love with the beautiful Indian princess and she fell in love with him.

Rolfe wrote to Governor Dale, asking to marry Pocahontas "to whom my heart and best thoughts are, and have a long time been so entangled, and enthralled in so intricate a labyrinth, that I was even wearied to unwind myself thereout."

It was a fine wedding, and it took place in the church at Jamestown. The Powhatan wouldn't come. Perhaps he feared a trap, or perhaps he was sad to see his daughter leave the Indian world. He sent two of her brothers and her uncle, Chief Opechancanough. They came wearing handsome garments of leather, furs, feathers, and beads. Pocahontas's marriage helped bring peace between Indians and colonists.

Soon Pocahontas had a baby and John Rolfe was so proud he took his family to England. There Pocahontas charmed everyone—even King James. (And King James didn't charm easily; he was a bit of a grouch.) The English people called her Lady Rebecca and treated her as the princess that she was. People fussed over her and pointed to her when she walked down the street. She was a celebrity. But she must have longed for a familiar face, someone she could talk to in her native language. And then she learned that John Smith was alive! She expected him to come and see her at once, but he didn't. She waited and waited. Finally, when he did come, she was so hurt that at first she wouldn't even talk to him. John Smith had been honored by her father. He was a member of her family. How could he ignore her? But Smith seemed different, and awkward. Perhaps he didn't know how to be her brother now that she wore a long dress like a proper Englishwoman. Perhaps Pocahontas herself didn't know in which world she belonged. When John Rolfe decided to go back to Virginia, she didn't want to go. She didn't feel at all well when they boarded a ship and sailed down the Thames river. Before they reached the open sea, she was so sick that her husband took her off the ship. It was smallpox. Pocahontas died, and was buried in the churchyard at Gravesend, a town that is now part of London, but in those days was in the country. She was 22.

English noblemen such as Lord de la Warr (above) often have two names: a title (such as Lord de la Warr) and their own given and family names. Lord de la Warr's name was Thomas West.

Let's take ourselves back in time to Jamestown. The Powhatan has not yet learned that his beloved daughter is dead. When he does, the peace will be finished. The Indians will attack. They will kill settlers, burn their homes, and try to drive them from the land. But now, for a while, there is calm.

8 A Share in America

Indian corn and an "Indian" jaybird illustrated a booklet advertising the New World's attractions.

More English ships sail for Jamestown...and more settlers...and more again. An English poet writes:

God will not let us fail.
Let England know our willingness,
For that our work is good;
We hope to plant a nation
Where none before hath stood.

Englishmen and women begin to spread out beyond Jamestown. They are settling in the New World.

It seems as if everyone in England wants to be part of the American adventure. And everyone can be part of it by giving money to the Virginia Company—the new name of the London Company—the outfit that is paying for all the exploration. The Virginia Company is a stock company, just like stock companies today. You can buy shares in the company; your money helps pay the company's expenses; if there are profits you will get your share of them.

Lords, knights, gentlemen, merchants, and plain citizens buy shares in the Virginia Company. The Archbishop of Canterbury, the Earl of Pembroke, famous Londoners, and unknown squires —all are among the investors.

The Spanish ambassador in London writes to King Philip that "fourteen earls and barons have given 40,000 ducats, the merchants give much more, and there is no poor little man nor woman who is not willing to subscribe something...much as I have written to your Majesty of the determination they have formed

How long would you last in this armor—on a humid summer's day in Virginia, out looking for deer?

THE
DISCOVERY
OF
Nevv Brittaine.
Began August 27. Anno Dom. 1650.
By Edward Bland, Merchant.
Abraham Woode, Captaine.
Sackford Brewster, Elias Pennant, Gentlemen.
From Fort Henry, at the head of Appamattuck River in Virginia, to the Fals of Blandina, first River in New Brittaine, which runneth West; being 120. Mile South-west, between 35. & 37. degrees, (a pleasant Country,) of temperate Ayre, and fertile Soyle.

LONDON,
Printed by Thomas Harper for John Stephenson, at the Sun below Ludgate. M.DC.LI.

The early promoters of the Virginia settlements had an uphill job, for everyone in England had heard about the terrible experiences of the first Jamestown colonists.

Class System

To some extent, the English will succeed in bringing their class society to Virginia. An upper class of landowning aristocrats will be the colony's leaders. There will be a middle class of yeomen owning small farms. The lower class will be made up of indentured servants and slaves. The Virginia aristocracy will differ from the English aristocracy in an important way: the idea that gentlemen should not work will be rejected in the New World.

here to go to Virginia, it seems to me that I still fall short of the reality."

Some of those people expect to make money from the gold that they are sure will be found. Many just want to take part in a great national venture. Some want to save the North American Indians from the Spaniards. In England people have read stories about the way some Spaniards treat the Indians: how they make them dig gold, how they starve them, how they make slaves of them. Good people are horrified.

They don't realize that English men and women are just like Spanish men and women. Some are good, some are not so good.

Some of the not-so-goods come to Jamestown. It is understandable that they would. Some people leave England because they are brave and curious; some are looking for riches or a better life. But others come because they are criminals and unwanted in England.

An Englishman named William Tucker arranges a powwow with the Pamunkey Indians. He tells them he wants to sign a peace treaty. The Indians give him corn and sign the treaty. Then Tucker suggests they all celebrate by drinking wine. He doesn't tell them that he has poisoned the wine: 200 Indians die.

Other Englishmen attack and burn Indian villages, sometimes for no apparent reason, sometimes to revenge the Starving Time.

Meanwhile, back in England, some people are writing of Virginia as a place where Native Americans and Europeans can live side by side and learn about the best of each other's culture. They talk of Indians going to English schools and Englishmen being trained in

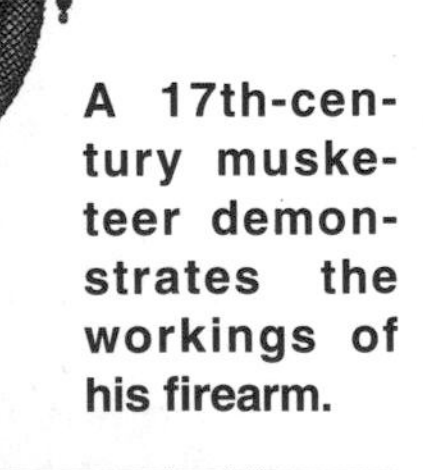

A 17th-century musketeer demonstrates the workings of his firearm.

Indian ways. An English school for the Indians is started at Henrico.

The people who have these fine ideas aren't concerned with gold. They picture an ideal nation where people live happily tending gardens and working at jobs like blowing glass, forging iron, and making perfume.

But the people with the good ideas stay in England. They are making plans for others, and that usually doesn't work. Those who come to Virginia aren't interested in ideal societies. Most have come for adventure, or to get rich, or to get away from their problems in Europe. After they get here, they find it a lot of trouble just trying to stay alive.

Of the first 10,000 settlers who land in Virginia, only 2,000 survive. (What percentage is that?) They die of disease, Indian attack, and hunger.

Hunger? In a land of plenty? Yes, even in a land of plenty you need to know how to hunt, or fish, or raise crops, or gather food. You need to adapt yourself to new conditions. If ever you go to a strange land, try and see how the natives live. You will learn a lot. That is what John Smith did. Some of the gentlemen do not learn. They try to be Englishmen in a wilderness.

Picture Jamestown in August. If there were a thermometer it would register 98°. The English gentlemen sometimes wear 60-pound suits of metal armor. That makes them feel safe from Indian arrows. Under the armor they wear wool clothing, because that is what they have always worn in England. Now, imagine chopping down a tree in that outfit: your hands are blistered—you've never held an ax before—and the biting flies are driving you crazy. Are you beginning to get angry? Look at the swampy ground. It's full of snakes and frogs and bugs—not the gold you were promised.

So you throw down your ax and decide to go hunting. Maybe you can kill a deer—you've seen plenty of animals. But that heavy musket is a problem. It makes a loud noise when it goes off, and you can't seem to hit anything with it, certainly not anything moving. A musket isn't accurate, like a rifle—but rifles haven't been invented yet. (A musket isn't as accurate as a bow and arrow, either.)

There is another problem I haven't mentioned. You stink. You haven't had a bath—ever. (Well, if you were daring, you may have had one last year.) Baths are considered unhealthy. Now remember, it is August and you are wearing all those clothes. Phew! If the deer aren't scared off by your loud, clanking armor, the smell will soon have them running. Indian hunters are quiet as falling leaves. They take off most of their clothes in summer; they bathe in the river. It is no wonder the Indians think the Englishmen are savages.

Tidewater **Virginia's diseases are new to the Europeans and kill many of them. But the Virginia diseases are not as deadly to the newcomers as the European diseases are to the Native Americans. Tribes are being wiped out.**

Wait a few **chapters and you'll see: there were fewer deaths in New England. The colder climate seemed to keep some germs under control, and the community there was better organized.**

9 Jamestown Makes It

In the words of the king of England himself, James I: "Smoking is a custom loathsome to the eye, hateful to the nose, harmful to the brain, dangerous to the lungs, and in the black, stinking fume thereof, nearest resembling the horrible Stygian smoke of the pit that is bottomless."

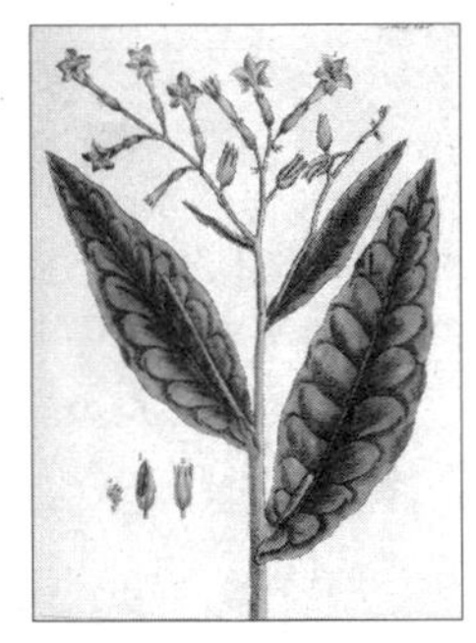

Taino Indians of the Bahamas grew tobacco; Spaniards probably copied its Taino name, *tabaco*.

At last the settlers found gold. Gold in the form of a leaf. A leaf that dried to a golden brown and could be put in a pipe and smoked. That tobacco leaf made men rich; it made the Virginia colony prosperous.

King James hated tobacco. He thought it unhealthy and he was right. But there is a limit to what even kings can do when money is involved. Growing tobacco was very profitable, especially after John Rolfe, Pocahontas's husband, developed a sweet variety that was all the rage in England.

But there was a problem. It takes hard fieldwork to grow tobacco, and Englishmen were not anxious to work in the fields. Besides that, even the best farmers could tend only a limited number of tobacco plants. So if you wanted to get rich by growing tobacco, you had to have people working for you. The more people you had, the more tobacco you could grow. The more tobacco you sold, the richer you would get. That made servants and other workers very valuable in Virginia.

So the Virginians did everything they could think of to get people to come to America. But since most of the settlers were dying, it wasn't easy. Most of those who came were poor or in trouble with the law.

The colonists were so eager to have workers that they were willing to pay for them. Sometimes they paid so much money that ship's captains would kidnap people from the streets of London.

Many of those who came to Virginia started out as indentured servants, and usually they were very poor. Some of them were criminals who were let out of jail if they would agree to come to the colony. You can understand that most people didn't want to go to a land where so many people were dying. The indentured servants didn't have enough

European tobacco merchants advertised their product with fanciful pictures of the exotic New World and its inhabitants.

money to pay their boat fare to the New World. They had to work for the person who paid the fare. They worked from four to seven years before they were free. That was their time of indenture. Some indentured servants were treated just like slaves.

What about slaves? Were there slaves in Jamestown? Yes, there were. Slavery in the English colonies began without much thought, which is the way bad things often begin.

In 1619 a Dutch ship brought a boatload of Africans to Jamestown. These people had been kidnapped from their homes by African traders and sold to the ship's captain. He in turn sold them to the Virginia settlers. Those first African Virginians were treated like indentured servants. After a few years of working for someone else, they became free. Soon there were Africans who had land of their own—and servants, too. But some colonists got the idea of making black people into slaves. That way they wouldn't have to keep buying workers on the docks. It must have seemed a good idea to people who were desperate for workers. Tobacco agriculture demanded much labor as well as a lot of land. There was an abundance of land in America, but few people willing to do hard work in the fields.

When Indians were enslaved, they ran away. It was difficult for the blacks to run away. Where would they go? Everything was new and strange to them. Gradually laws were passed to trap black people in slavery. It was the beginning of a way of life that would bring misery to many, many innocent African Americans.

Why did Europeans go to the trouble of importing African slaves instead of forcing Native Americans to work for them? Because the Indians didn't make good slaves. They got sick from Old World diseases, and often they just ran away.

How Tobacco Beat Out Silk

The silkworm is the caterpillar stage of the silk moth. The material that the silkworms spin into their cocoons is what makes the silk thread.

King James wanted to start a silk industry in Virginia, and the colonists needed a doctor. Dr. Lawrence Bohune was the perfect man for both tasks: he was a physician and he had scientific curiosity. He planned to experiment with silkworms and also to investigate the native herbs and plants the Indians used for healing purposes. Bohune had visited the colony in 1610 as Lord de la Warr's personal doctor, and he had impressed everyone with his good sense. So when he made plans to bring silkworms to Virginia, the king and the settlers were pleased.

In 1620, Dr. Bohune set sail for Jamestown on a ship named the *Margaret and John.* (It was the very year that the Pilgrims arrived in Massachusetts Bay.) After 11 tough weeks at sea, the small ship reached the West Indies and found itself facing two armed Spanish warships. The Spaniards fired their cannons. The English ship was outclassed, a cannonball struck the good doctor, and he fell into the arms of the captain. "Oh, Dr. Bohune, what a disaster this is!" said the captain. With his last breath the doctor replied, "Fight it out, brave man, the cause is good, and the Lord receive my soul."

When the damaged ship limped into Jamestown, the doctor was dead and the silkworms had all been lost at sea. King James had hoped that silk would replace tobacco as Virginia's gold. It never did.

10 1619— A Big Year

If the colony was to survive, it had to grow. That meant sending women as well as men.

The English found those first years in America really hard. Remember, four out of five of the first 10,000 settlers died soon after they arrived in Virginia. Most people would have given up—but not the English. The harder the challenge, the more determined they became.

The year 1619 was a turning point. After 1619 you could tell the English were in America to stay. It was a year of many firsts in Virginia:

- first boatload of Africans
- first boatload of women
- first labor strike
- first time English settlers are allowed to own land
- first elected lawmakers.

That is a lot for any year. You already know about that boatload of Africans. Now, about those women. They, too, were sold on the docks.

"Do you want a wife?"

"It will cost you 120 pounds of tobacco."

Those are the terms when a shipload of women arrives in Jamestown in 1619. These are poor women who are unable to pay the cost of their Atlantic journey. They want a new life in this new land. The lonely men want wives. There will be instant romances on the docks. What do you think of these women? Do you think they are scared? Courageous? Crazy?

A few white women have already been to Jamestown, but sending an entire boatload of them to be wives means that the English plan to

Some of the early laws passed by the burgesses of Virginia forbade pastimes that were thought immoral, like cardplaying and throwing dice. If you got caught not going to church, you were fined 50 pounds of tobacco—about a week's wages. It was against the law to swear, too.

stay and make homes in America. The French, who are settling in the North, are less likely to send women. Still, in 1619, the men in Jamestown outnumber the women by eight to one.

Some historians think the reason there was much violence in Jamestown was because many more men than women and children lived there. Do you think that is true? Do you think men living alone fight more than people in families? That is something to discuss.

In 1619 the Virginia Company lets the settlers have land of their own. That gives them a reason to work hard.

The first workers' strike in British America happens in 1619. Polish workers at Jamestown, who are glassmakers, demand the same rights as Englishmen. They get those rights and go back to work. There are Poles, Dutch, Germans, and Italians at Jamestown. Do you think it strange that they all want English rights? What about their own rights?

The answer to that is very simple. English men and women have more rights and freedom than people do in other European nations. They expect those same rights in America and so do people from other nations who come to the English colonies.

After seven years' work, those who wanted were given their own land. Captain Smith said, "When our people were fed out of the common store, and laboured jointly together, glad was he who could slip from his labour, or slumber over his tasks, he cared not how; nay, the most honest among them would hardly take so much true paines in a week, as now for themselves they will do in a day."

John Smith said no one would come to the New World "to have less freedom." The Charter of the Virginia Company said, "all and every of the persons...which shall dwell and inhabit within every or any of the said several colonies and plantations, and every of their children... shall have and enjoy all liberties...as if they had been abiding and born, within this our realm of England." That means that nobody will lose freedom if he moves from England to America.

In 1619, a group of lawmakers—known as burgesses—is elected to make laws. They form an assembly called the House of Burgesses. In England laws are made by Parliament. The House of Burgesses gives the Virginians their own form of Parliament. That has never happened in a colony before.

Once a woman got to the New World, she was expected to start having babies—that's what the Puritans on the left are telling the couple to do.

By the way, do you know what a colony is?

A colony is land controlled by a distant, or foreign, nation. In the 17th century many European nations have colonies in America as well as in other parts of the world. Those colonies are not all alike.

In the Spanish colonies no Europeans except Spaniards are allowed to settle. France admits only Catholics. The English colonies have open doors.

Think about that for a minute. That decision, way back in the 1600s, to let all kinds of people settle in the English colonies, made a big dif-

A ***realm*** is the kingdom or country where a ruler holds sway.

A Hostage Swap

In 1611, Captain John Clark was a pilot on one of three ships bound for Virginia. (Like an airplane pilot, a ship's pilot is expected to guide his craft safely.) Clark and his ships (Christopher Newport was in charge of the expedition) made it to Virginia. They hadn't been there long when a Spanish ship sailed into Chesapeake Bay. A Spanish officer came ashore and the English took him hostage. Then the Spaniards captured Clark and sailed off to Spain. Five years later the two prisoners were exchanged in London. In 1620, the Virginia Company asked John Clark to pilot a small ship to Virginia. It was the *Mayflower*, and it didn't quite reach its destination.

ference to our country. We would become a pluralistic society. (What does that mean?)

Now that you know about colonies, let's get back to the House of Burgesses. In the 17th century, laws for colonies were made in the home country, or by appointed governors and their councils. The House of Burgesses changed that.

England was letting colonists make laws for themselves. That was a big first in history. (An English governor did have *veto power* over the burgesses. What is a veto? Okay, you can do some work. Go to the dictionary and look that word up. The governor didn't use the veto very often.)

This is something you should remember: the House of Burgesses, formed in 1619, gave America its first representative government. It was the beginning of self-government in America.

Whoops! Hold on, that isn't quite true. Some Indian tribes had representative government. The House of Burgesses was the first representative assembly in the European colonies.

It was only a dozen years since those three small ships were tied to the trees at Jamestown and the English colonists were doing something very unusual. They were making laws for themselves.

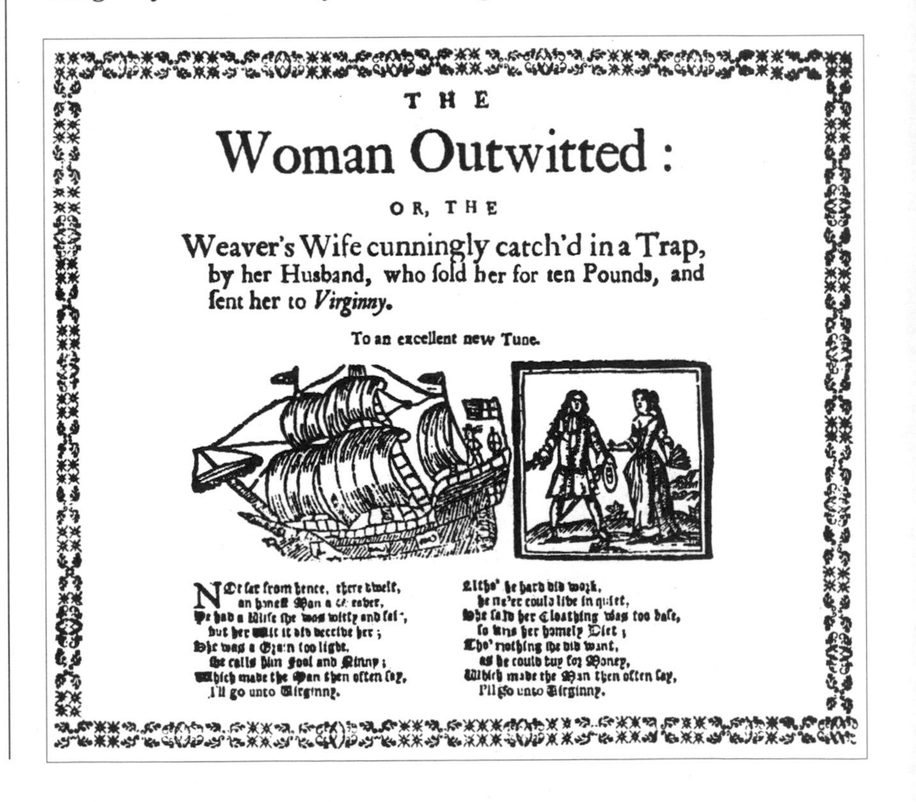
THE

Woman Outwitted:

OR, THE

Weaver's Wife cunningly catch'd in a Trap, by her Husband, who sold her for ten Pounds, and sent her to *Virginny*.

To an excellent new Tune.

Not far from hence, there dwelt,
an honest Man a Weaver,
He had a Wife she was witty and fair,
but her Wit it did deceive her;
She was a Grain too light,
she calls him Fool and Ninny;
Which made the Man then often say,
I'll go unto Virginny.

Altho' he hard did work,
he ne'er could live in quiet,
She said her Cloathing was too base,
so was her homely Diet;
Tho' nothing she did want,
as he could buy for Money,
Which made the Man then often say,
I'll go unto Virginny.

The Virginia Company was so eager to get women to come to America that it sometimes resorted to buying girls from their families or even kidnapping them from their homes.

11 Indians vs. Colonists

The Powhatan Indians played football with a small ball and a goal. The men played one set of rules, the women and children another. But for all the players, speed and dexterity were important. The Indians also enjoyed a game called *chunkey*, which was played with a disk-shaped stone and slim poles eight or ten feet long. The idea was to roll the disk as far as possible and then throw the pole and try to hit it—or knock your opponent's pole away from its target disk.

If the European settlements were to grow, they had to have the Indians' land. And the Indians weren't about to give it up easily.

From the time of Columbus the pattern was the same. The newcomers and the Indians would meet as friends and trade with each other. Then something would happen. Often an Indian was killed or sold into slavery, and the Indians would strike back. Sometimes they showed remarkable patience. Sometimes they were just waiting for the right moment. For the Native Americans were much like the New Americans: good and bad, fierce and gentle. Warriors on both sides went too far. The massacres were horrible.

At first the Indian leaders tried to live in peace with the settlers. But some of them realized that it would not work, that it would be the end of Indian ways. The Europeans used up land. They cut the forests and filled the land with people. Indians were hunters. To keep their way of life, the woods had to be protected. Wild animals need woods to live in, and hunters need wild animals.

Most Europeans understood that, too. One Virginia governor said, "Either we must clear the Indians out of the country, or they must clear us out." The members of the House of Burgesses ordered three expeditions to drive out the Indians "in order that they have no chance to harvest their crops or rebuild their wigwams."

There was another problem: arrogance (which means thinking you are better than others). In the 17th century arrogance was usually tied to religion.

What do you think about this Native American cartoonist's version of the Pilgrims' Thanksgiving story?

Fields of Blood

Not all white people feared and hated Indians. John Lawson, a traveler in Carolina in 1700, wrote in a book about his journey: "We look upon them with Scorn and Disdain, [yet]...for all our Religion and Education, we possess more Moral Deformities and Evils than these savages do....We make way for a Christian Colony through a Field of Blood, and defraud, and make away with those that one day may be wanted in this world." *John Lawson founded a new settlement himself. He was later captured and killed by Tuscarora Indians.*

Many Christians believed that anyone who was not Christian must be inferior. (The Aztecs believed those who weren't Aztec were inferior.)

Before long, that arrogance would become racism. Some whites believed themselves better than all Indians. Some believed themselves better than all blacks. History shows that racists are troublemakers and often the worst of their own race. There were bigots and racists in early America and they made trouble. Some of them wanted to kill all the Indians. (And some Indians wanted to kill all whites.)

However, the real problem was the fight for control of land. Even when Indians and settlers were friendly, it usually didn't last long. The newcomers wanted Indian land, and naturally the Indians didn't want to give it up. Some fair-minded white leaders respected the Indians and wanted to share the land, but they were never able to control the land-hungry settlers.

In 1737 the Delaware Indian chief Lappawinsoe signed an agreement that gave Pennsylvania colonists all the land they could cover on foot in a day and a half. Instead of walking leisurely, as the Indians expected, the colonists sent runners. Lappawinsoe was shocked and felt the Indians had been swindled.

Colonists confront Indians during King Philip's War. Bows and arrows were faster, but guns seemed scary and noisy.

12 Massacre in Virginia, Poverty in England

Opechancanough was a sachem, an important Indian chief and wise man.

Just when things seemed to be going well for the colonists, Pocahontas's uncle, the sachem Opechancanough, decided to try to get rid of them all. Some historians think that Opechancanough was the Indian prince who had been taken to Spain by Spanish priests, educated, and returned to his people. Whether he was or not, everyone agrees that he was intelligent and crafty, and that he hated the white men who were stealing his land.

In 1622 Opechancanough was an old man. Perhaps he thought it was his last chance to save his people. So he planned a great massacre. Indians knocked on the colonists' doors—pretending to be friendly—and then they murdered and scalped. They might have killed everyone if an Indian boy, Chanco, hadn't warned the men and women at Jamestown. Chanco had been treated kindly by the settlers and had become a Christian. The settlements outside Jamestown didn't get warned. Hundreds of English men, women, and children were killed in the Great Massacre of 1622.

King James was upset; there were too many deaths in Virginia. He set up a government investigation, and then he closed down the Virginia Company. The stockholders were wiped out; their stock was now worthless. King James took Virginia; it became a royal colony. Actually, the king didn't spend much time thinking about the Virginia colony; it was too far away. And he had problems, big problems, at home.

James believed that God had given him the right to rule—he called it divine right. He thought that divine right meant he could do almost anything he wanted to. Parliament didn't agree, and Parliament controlled

Opechancanough never gave up. In 1644 his warriors attacked Jamestown again. He was so old and feeble he had to be carried about on a bed. Attendants held his eyelids open so he could see. But his mind had not lost its power.

To ***massacre*** (MASS-uh-ker) means to kill brutally and often in large numbers.

After the Jamestown Massacre the English had an excuse for killing Indians, and the bloodshed became intense. Killings were followed by revenge raids, more killings, and more revenge—on both sides.

In what languages was the Bible first written? See if you can find the answer.

To ***dispute*** means to argue. To ***harangue*** means to talk or lecture someone very forcefully. ***Drivel*** is stupid talk.

Bibles and Books

The King James Bible is made up of two parts: the Old Testament and the New Testament. These books, or parts of them, have other names too. The Old Testament is also the *Hebrew Bible*; and the first five books (do you know their names?) in the Old Testament make up the Jewish *Torah*. The Greek name for the first five books is the *Pentateuch* (PEN-tuh-tewk). The first four books of the New Testament are called the *Gospels*. Another name for the King James Bible is the "Authorized Version"—because its publication was authorized by the king himself.

most of the money in England. Parliament wouldn't give James the cash he wanted. Things got edgy.

James was a thoughtful man who might have made a fine professor. While he was on the throne the Bible was translated into English. That translation is called the "King James Bible." It was read in most Protestant churches in America until the 20th century. Many people think it the most beautiful translation ever.

But what England needed was a strong political leader, not a professor. One historian said King James was "two men—a witty, well-read scholar who wrote, disputed and harangued, and a nervous, drivelling idiot who acted." King Henry IV of France called him "the wisest fool in Christendom." James just wasn't the right type to be a king. He was in the wrong profession.

Let's get into a time capsule and take a look at King James's realm. Things are not going well at all. Farmers, who rent land from the rich landowning lords, are being thrown off their farms. The landlords want the land because of the new craze for sheep raising.

For some reason no one quite understands, the population is growing faster than it has ever done before. Jobs are hard to find. London and the countryside are full of beggars and starving people. Some of them climb on ships and pray for luck and a better life in the New World. Boatloads of people begin crossing the ocean.

Many of those who sail are convicts let out of jail if they will make the voyage. Englishmen write of America as a place to send their poor and troubled.

Some of the settlers are orphans. Many are very young. Heat and germs and Indians will kill most of them, yet they keep coming.

By midcentury (which century?) there are brick houses at Jamestown, a brick church, a fine State House, and plenty of food. For the European settlers, the American Dream has begun. Those who are tough and work hard will find in America a land of opportunity, like no land before it.

Opechancanough and his warriors attacked Jamestown on March 22, 1622. They killed nearly a third of the town's 1,200 inhabitants. It wasn't as one-sided as you might think from the picture, though. Afterwards, the settlers destroyed many of the Indians' villages, and their crops, too.

America: Land of the Free

From its beginnings, America was a land of freedom and opportunity for all. True or false?

The answer is FALSE.

For many of us, America was a land of humiliation and enslavement.

Africans came to the New World not because they wanted to but because they were taken from their homes by men with powerful weapons. When they protested, they were beaten and killed.

There was big money in it for those who stole them. There was big money in it for those who transported them. There were profits and an easy life for those who bought them.

Slavery had been around for a long time when Columbus set sail. When Moses led the Jews from Egypt, they were escaping slavery. The ancient Greeks and Romans kept slaves. But in olden days, slaves were usually the booty of war. If you were captured in battle, you might end up a slave of the enemy. Slavery had nothing to do with skin color. Slaves were sometimes allowed to buy their freedom. Children of slaves were not always enslaved.

It was a Portuguese prince who got the African slave trade started in Europe. Remember Prince Henry the Navigator? In 1442 one of his ships arrived in Portugal with ten captured Africans. The Portuguese were looking for ways to make money. Africans were good workers. Selling them as slaves would be a profitable business. Prince Henry gave those ten black Africans as a gift to the Roman Catholic Pope. The Pope gave Portugal the right to trade in Africa. But by 1455 the slave trade had become so abusive that Prince Henry tried to stop it. It was too late.

When the New World was discovered, workers were needed to mine its resources and to work its fields. Europeans didn't want those jobs; slaves had no choice. Slavery in America developed into a terrible and degrading system, worse than any in previous history. To justify that terrible system, a myth arose that blacks were inferior, that they weren't capable people.

Of course, that was just a myth. Africa had produced great cities and beautiful arts. In the 11th century the great African empire of Ghana was flourishing. Al-Bakri, a geographer living in Spain, wrote of a Ghanaian city with "fine houses and solid buildings" and of a royal pavilion where pages held gold-tipped swords, horses gleamed in cloths of gold, and princes were "splendidly clad and with gold plaited into their hair." In the 15th century, scholars came from Europe to study history, medicine, law, and literature at Sankore in Timbuktu, a Muslim center of learning in West Africa.

In America black people worked hard, and an entire way of life depended on their labor. Despite their harsh treatment, they produced writers, scientists, political leaders, musicians and many others who enriched our nation.

A famous philosopher named George Santayana said, "Those who cannot remember the past are condemned to repeat it." What do you think he meant? Do you think he was talking of nations, or people, or both?

One thing he didn't mean was that the same mistakes are made again and again. We will never again allow the kind of slavery we had when this nation began. But there are other forms of slavery. What about drugs? Do they enslave people's minds and bodies? Do they kill? Why do people sell drugs?

In the 20th century slave-labor camps have been the scene of brutality and death for millions of Europeans and Asians. Do you know about them?

Do you know about apartheid (uh-PAR-tide) in South Africa?

American slavery was a horror. We should never pretend it was anything else. But the American system of government lets us correct our mistakes. When you study history you see that we usually do. Of that we can all be proud.

13 The Mayflower: Saints and Strangers

The plaque at St. Peter's Church in Leyden, Holland, put up in memory of the *Mayflower* Pilgrims who once worshiped there.

The times were religious—and angry. To understand them we need to review some English history. Remember King Henry VIII? He was the father of Queen Elizabeth. King Henry tossed the Catholic church out of England long before Jamestown got started. Why Henry did that is an interesting story, but you'll have to look up the details yourself. It had something to do with King Henry's wanting to get married again, and again, and again, and—whew—he had a lot of energy.

The head of the Catholic church, the Pope, didn't approve of all that marrying. So King Henry founded the Church of England (which is sometimes called the Anglican church) and made himself its leader.

By the 17th century most English men and women belonged to that church. (As they still do.) It was called the "established church" because it was linked to the government. Since he was king, James was head of the Church of England. The man who actually ran the church was called the Archbishop of Canterbury, and he was appointed by the king.

The Pope lived in Rome in a great palace called the Vatican. That was, and is, the control center for the entire Roman Catholic church. The Pope was elected by bishops of the Catholic church.

Except for that matter of control and leadership, the Anglicans and Catholics were much alike, although they didn't think so and often hated and persecuted each other. As we said, the times were not only

In December 1620 the *Mayflower* was in Cape Cod harbor. Before the Pilgrims disembarked for good at Plymouth, Mistress Susanna White gave birth on board the ship to the first English baby born in New England, a boy whom she named Peregrine (PAIR-uh-grin). Here is his cradle.

John Smith offered to hire himself out to the Pilgrims as their guide. They told him his book was "better cheap" than he was.

Changing Times

James I brought "new ideas, new goings, new measures, new paces, new heads for your men, for women new faces," according to a poem of the times. It was a time of increasing lawlessness, of new riches for some, of poverty for others. There was growing interest in individual rights; less in community values. People began to focus on material riches rather than religion. All that was confusing and disturbing to many. Have you heard people you know complain about politics and government and the times? The Pilgrims didn't like the politics of their England, so they left.

religious, but also intolerant. People took their differences very seriously. Wars were fought over them.

Some Englishmen wanted the differences between Catholics and Protestants to be greater. They felt that King Henry VIII didn't go far enough when he outlawed the Catholic church. They didn't want the Anglican church service to be at all like the Catholic service. They said they wanted to "purify" the Church of England, so they were called Puritans.

Others wanted to go even further. They believed people could speak directly to God without a priest or bishop at all. They wanted to separate themselves from the Church of England and form congregations of their own. They called themselves Saints. Other people called them Separatists. Some people called them troublemakers.

King James would not let the Separatists practice their religion. They had to go to the Church of England or go to jail. Their religion was more important to them than their homes—and sometimes than life itself. Some of the Separatists, especially a group from a village in northeast England called Scrooby, decided to move to Holland, where they were promised religious freedom.

And they got religious freedom in Holland—but they didn't feel at home with the Dutch. They were English, and they liked their own customs and language and villages. When their children started speaking Dutch and forgetting English ways, the people from Scrooby decided it was time to move again. They read John Smith's book, *Description of New England*, and they said, "This time to America."

Anyone who takes a trip for religious purposes is a *pilgrim*. So now these Scrooby people who were called Separatists or Saints had a new name: Pilgrims. They were the first of many, many boatloads of pilgrims who would come to America to be free to

The Pilgrims in Holland board ship at Leyden for the first leg of their journey to religious freedom in the New World.

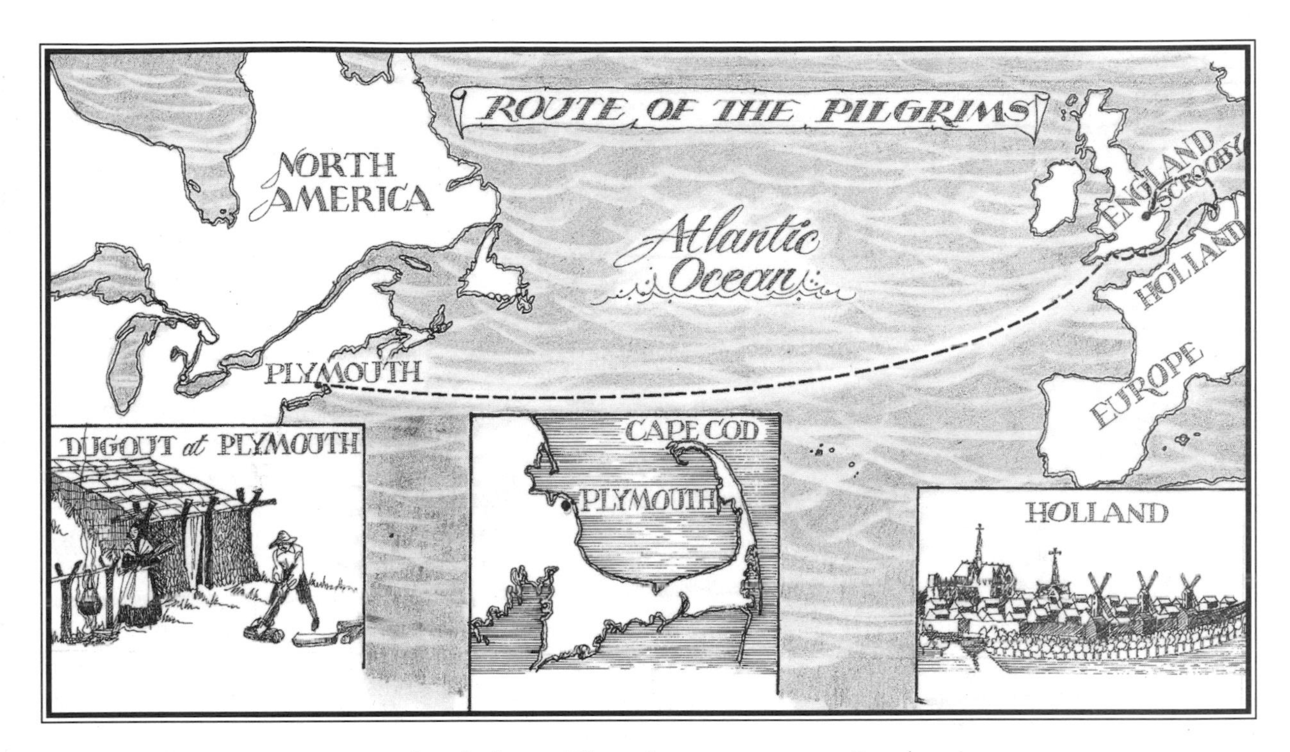

believe whatever they wanted to believe. They, however, were pilgrims with a capital P: *the* Pilgrims.

The year is 1620. The boat they take is named the *Mayflower*. Of the 102 on board, only about half are Saints; the Scrooby people call the others Strangers. The Strangers are leaving England for adventure, or because they are unhappy or in trouble. Saints and Strangers have many things in common. Most are from the lower classes; most have a trade; they expect to work hard; they are ambitious; and they can't stand the new ideas that are changing England. All want a better life, but the Saints hope to build a society more perfect than any on earth.

Among the Strangers are 10 indentured servants, a professional soldier, a barrelmaker, four orphans indentured, or "bound," to work without pay until they are 21, and a man soon to be convicted of murder.

Among the Saints is William Bradford, who will be elected as the colony's second governor and will write their story.

It is a terrible voyage, taking 66 days. The ship is small, wet, and foul. The smells are horrid. There is no place to change or wash clothes. Each adult is assigned a space below deck measuring seven by two and a half feet. Children get even less room. None of the passengers is allowed on deck; there is little fresh air below and many are sick. Fresh food soon runs out and then there is hard bread and dried meat that is wet and moldy. But the Pilgrims have onions, lemon juice, and beer to keep them from getting the dreaded scurvy. Amazingly,

The word *indentured* originally came from the paper that the contract between master and servant was written on. After they signed it, the paper was torn in half, so that each piece had an *indentation* that fitted into the other piece. The master kept one piece and the servant kept the other. That was the proof of their agreement.

Scurvy is a disease resulting from lack of vitamin C. It makes people bleed easily and causes their teeth to fall out.

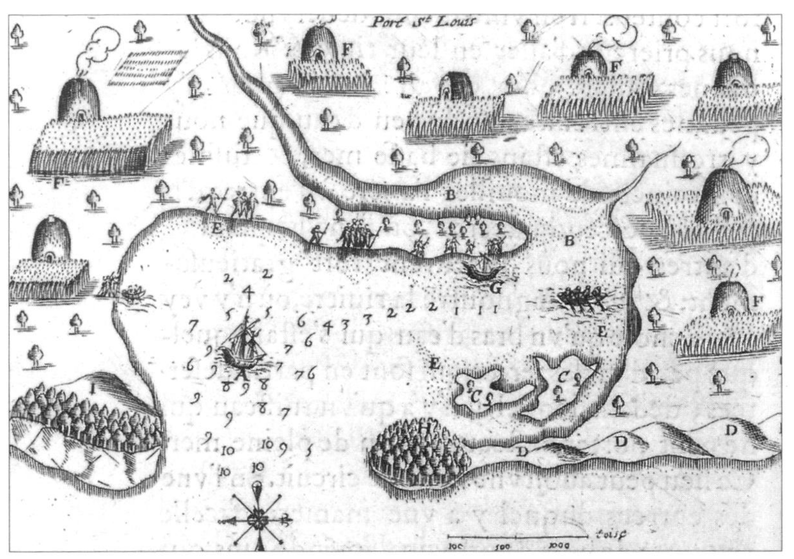

The Pilgrims weren't the first Europeans to land at Plymouth Bay. Fishermen, trappers, and explorers, such as the Frenchman Samuel de Champlain, had been there before. Champlain drew this map, showing the Patuxet Indians' villages, in 1605.

only one person dies. He is replaced on the roster by a baby born at sea, Oceanus (oh-shee-ANN-us) Hopkins. Another child, Peregrine White, is born just before they dock.

When they first sight American land, it is at Cape Cod. They planned to go to Virginia, but they are exhausted. Bradford describes Cape Cod as a "hideous and desolate wilderness, full of wild beasts and men." They sail around the Cape to a place they see on Smith's map. He has called it Plymouth, after a town in England.

Before they get off the ship, there are matters to attend to. There has been trouble between Saints and Strangers, and it needs to be settled. They must live together peacefully. They need rules and laws and leaders. So they draw up a plan of government, the Mayflower Compact, which establishes a governing body:

> *to enact, constitute, and frame such just and equal laws, ordinances, acts, constitutions, offices...for the general good of the Colony; unto which, we promise all due submission and obedience.*

That Mayflower Compact is one of the great documents of American history. Here is a group of settlers able to govern themselves; reasonable people who agree to live together under a government of laws. The king doesn't realize what is in the future. This breed of people will not allow others to rule them for long.

Then, wrote Bradford, "being thus arrived in a good harbor, and brought safe to land, they fell upon their knees and blessed the God of Heaven who had brought them over the vast and furious ocean."

When they land, they find empty fields cleared for planting. They will learn that smallpox, caught from white fishermen, has wiped out many of New England's Indians. The Pilgrims believe that God has made the land theirs for the taking.

But it is December—too late to plant crops. Many will hunger and die before spring comes. Fewer than half of the 102 who land will survive the first winter. But no one wants to return to England. These are sturdy folk who intend to start a nation. William Bradford writes of the colony "as one small candle may light a thousand, so the light here kindled hath shone unto many, yea in some sort to our whole nation."

Although the *Mayflower* was cramped and uncomfortable, one thing about it was an improvement over many ships. The *Mayflower* had carried wine barrels, and the hold smelled quite pleasantly of the wine that had leaked out. Most 17th-century ships stank of garbage.

14 Pilgrims, Indians, and Puritans

One way of cooking a turkey on an open fire was in a roasting pan with a tight lid: "For turkey braised, the Lord be praised."

Like the Jamestown colonists, the Pilgrims have picked a poor site. The New England coast is cold and wind-whipped; the land is rocky, the soil is thin. But these industrious people will use the sea and the forest to sustain themselves. Soon they will be shipping fish, furs, and lumber back to England.

Without the Indians they might not have survived. Picture this scene: Pilgrims are struggling to find ways to live in this difficult region, when out of the woods strides a tall man in deerskin clothes. They are astounded when he greets them. "Welcome, Englishmen," he says. His name is Samoset, and he has learned some English from fishermen and traders.

Samoset returns with 60 Indians, a chieftain named Massasoit, and an Indian whom the settlers name Squanto. Squanto speaks English well. He had been kidnapped by sailors, taken to London, befriended by a London merchant, and returned to his native land.

Trumpet and drums are played as the Pilgrims' governor, John Carver, leads Massasoit to his house, kisses his hand (as is proper to a king), offers refreshments, and writes a treaty of peace between the Indians and the English. While Massasoit is alive, the peace will be kept.

Squanto stays with the settlers. To the Pilgrims he was "a special instrument sent of God for their good beyond their expectation...He directed them how to set [plant] their corn, where to take [catch] fish, and to procure other commodities, and was also their pilot to bring them to unknown places."

In 1621, after the first harvest, the Pilgrims invite their Indian

Squanto's real name is Tisquantum. These Native Americans are Algonquians of the Wampanoag tribe, who live in what is now Rhode Island. *Wampanoag* means "eastern people." They hunt, fish, dig for clams, and gather berries and nuts. They are good cooks; they make venison (deer) steak, fish chowders, succotash, cornbread, and maple sugar.

Edward Winslow was a printer and a clever man. He traveled a lot to London to trade and negotiate for the Pilgrims. On his first trip back to Plymouth he brought something very important: cattle.

Not all the early colonists dressed in somber styles and colors. Only the Saints insisted on plain, dark clothes.

friends to a three-day feast of Thanksgiving. It is not the first Thanksgiving in America, but it is special. The Indians bring five deer; the settlers provide turkeys and other good foods. In one year they have accomplished much.

The Pilgrims are frugal, but the celebration is unusually generous. They will need their food to get through the winter and to help feed the new colonists who are beginning to arrive.

William Bradford, who is elected governor when John Carver dies suddenly, keeps a record of the arrivals. When the ship *Fortune* docks he writes, "there was not so much as biscuit-cake, neither had they any bedding...nor pot, or pan."

Abraham Pearce, a black indentured servant, is one of those who comes in 1623. A few years later he will own land, vote, and be a respected member of the community.

The new arrivals bring reports from England that are not good. Now the Puritans are in trouble. The Puritans are also called Saints, but they are more moderate than the Pilgrims. Remember, Puritans don't want to separate themselves from the Church of England, they want to purify the church. What they really want is to control the Church of England; of course, King James and those in charge don't want that at all.

The Puritans can't stand King James and he doesn't like them either. Of the Puritans he says, "I will make them conform themselves, or else I will harry [harass] them out of the land." The Puritans can see that King James isn't good for the economy. First there is inflation and then a depression. James has brought his big-spending friends to England from Scotland, where he is also king. They are getting special favors. The Puritans are not.

And so a group of Puritans gathers at Cambridge University, where most have gone to college, and makes plans to sail to America. The Puritans are better educated than the Pilgrims—and richer, too. John Winthrop, their leader, is a lawyer, born on a manor, with servants and tenants.

When King James dies, in 1625, and Charles I becomes king, things become even worse for the Puritans.

One of America's first folk songs praised the humble but essential pumpkin.

For pottage and puddings
and custards and pies
Our pumpkins and parsnips
are common supplies.
We have pumpkin at morning and pumpkin at noon,
If it were not for pumpkin,
We should be undone.

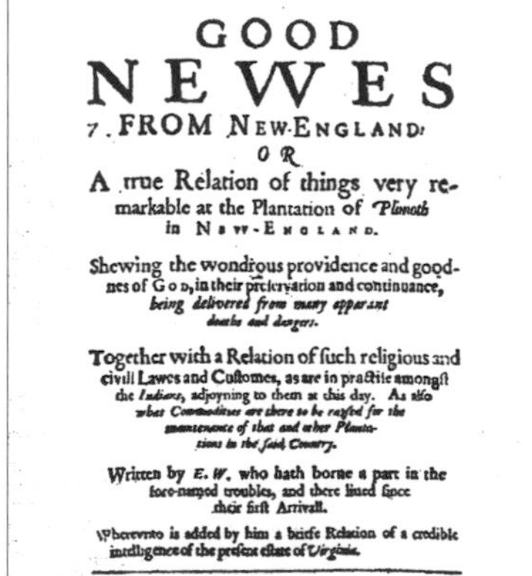

GOOD
NEWES
FROM NEW-ENGLAND:
OR
A true Relation of things very remarkable at the Plantation of *Plimoth* in NEW-ENGLAND.

Shewing the wondrous providence and goodnes of GOD, in their preseruation and continuance, *being delivered from many apparant deaths and dangers.*

Together with a Relation of such religious and civill Lawes and Customes, as are in practise amongst the *Indians*, adjoyning to them at this day. As also *what Commodities are there to be raysed for the maintenance of that and other Plantations in the said Country.*

Written by *E. W.* who hath borne a part in the fore-named troubles, and there liued since their first Arrivall.

Whereunto is added by him a briefe Relation of a credible intelligence of the present estate of *Virginia.*

LONDON
Printed by *I. D.* for *William Bladen* and *John Bellamie*, and are to be sold at their Shops, at the *Bible* in *Pauls*-Church-yard, and at the three Golden Lyons in Corn-hill, neere the *Royall Exchange*. 1624.

Edward Winslow wrote this pamphlet to encourage more colonists to come to New England.

15 Puritans, Puritans, and More Puritans

Massachusetts is an Algonquian word that means "at the big hill." The Puritans called their new home the Massachusetts Bay Colony.

John Winthrop was a lawyer who lost his job because of his religion. In 1630 he came to America.

In 1630, the first Puritan ship, the *Arbella*, sets out for the New World. By summer's end 1,000 Puritans have landed in New England. They bring a charter from the king: the Charter of the Company of the Massachusetts Bay in New England. King Charles is happy to see the Puritans leave England.

The charter is a document written by lawyers, setting the rules that tell how the colony will be run. It allows the colonists to govern themselves. It is important to remember that, from the beginning, English settlers expected to govern themselves. It is important to remember that each colony had a charter: a written set of rules. Those charters would evolve into constitutions.

Can you guess what might happen in a community without a charter or constitution? Would you like to live in a country without laws? Would you want to write your own laws or have someone write them for you?

You can think about those questions and then get back to the Puritans, who are beginning to pour out of England. Most of them go to the Caribbean islands, where sugar is creating great fortunes. But, between 1630 and 1640, 20,000 Puritans sail for New England. Think about all those people risking their lives to cross the ocean and settle in an unknown land. It is almost as if tens of thousands of people today decided to live in outer space.

Why did they come? Many came because they really cared about their religion and wanted to practice it in peace. They wanted to build a holy community, where people would live by the rules of the Bible. Puritans believed that the Bible was the whole word of God. They tried to follow its every direction, which means they tried to live very good lives.

Sweet Success

Sugar was a much-desired luxury in Europe. The West Indies (with its tropical climate) turned out to be a perfect place to grow sugarcane. Christopher Columbus brought the cane to the Caribbean on his second voyage. Soon the native trees were cut down and sugarcane plantations filled the islands.

Sugarcane grew wild in Asia. In Sanskrit, the ancient language of India, sugar was *sarkara*. That became *sukkar* in Arabic, *sakhar* in Russian, *sucre* in French, *Zucker* in German, and "sugar" in English.

Here is the first page of the Charter of the Massachusetts Bay Colony, setting out the rules by which the colonists could govern themselves. At the top left is King William III.

Although the Puritans tried hard to be good, their colony didn't work out as they wished. But how many people do you know willing to devote their lives to an idea they think is right?

They expected their colony to be an example for all the world. John Winthrop, who was chosen governor said, "We must consider that we shall be as a city upon a hill. The eyes of all people are upon us."

One thing they didn't understand at all was the idea of *toleration*. Puritans came to America to find religious freedom—but only for themselves. They didn't believe in the kind of religious freedom we have today. But don't be too hard on them. Almost no one else believed in it either.

In those days each nation had its own church, and everyone was expected to pay taxes for its support. Suppose you didn't believe in the ideas of that religion. Too bad. You had to keep quiet, leave the country, go to jail, or maybe get hanged.

Pretend you are a Puritan. You think that yours is the only true religion, so you believe the Reverend John Cotton when he says toleration is "liberty...to tell lies in the name of the Lord."

Since you are convinced that only you Puritans are right, you think it is wrong to let anyone practice another religion. You believe that is helping the devil. You especially dislike Quakers. Your leaders call them a "cursed sect." You use the name Quaker to describe religious people who call themselves Friends. Friends believe that each person has an inner light that leads him to God. People with an inner light do not have to rely on a minister to tell them what is godly. The inner light is available to everyone. This is a highly democratic idea, and most Europeans thought it very dangerous. They were used to kings and priests and ministers. It seemed reasonable to them to persecute Quakers. When Quakers came to New England or Virginia, they were whipped, sent away, and even hanged.

Remember, you are a Puritan and you've left your home and everything you know and love. You've crossed a fierce ocean to live as you wish. You don't want people with strange ideas bothering you. Democracy is another strange idea. "If the people be governors, who shall be governed?" the Reverend Cotton asks. John Winthrop, the beloved Puritan governor, who always tries to do what is best, calls

At first, the name *Quaker* was used to make fun of people; so too was the word *Puritan*. Then both groups decided to be proud of those words and use them themselves.

At this Quaker meeting most of the congregation sits in silent meditation while one member is inspired to speak.

democracy the "meanest [lowest] and worst" form of government.

And yet the Puritans do practice a kind of democracy—but only for male church members. Once a year they form a General Court and vote to elect the governor and council. The General Court is a lot like the House of Burgesses, or Parliament, or Congress.

Some people call the Massachusetts Bay Colony a *theocracy* (thee-OCK-ruh-see), government by church officials in the name of God. But they are wrong. It is not a theocracy. The ministers are the most important people in the colony, but they are not allowed to hold political office. They do not govern. It is a small step toward the idea of the separation of church and state. Someday that idea will be a foundation of American liberty.

What's in a Word?

We had better stop and go over some words, otherwise this book will get confusing. You've been reading about democracy and communism and theocracy. Do you know what they mean? Let's make sure you do.

If you keep reading history you will learn about Abraham Lincoln. He said that democracy was government "of the people, by the people, and for the people." That's a good definition.

Democracy is based on people power. Think of a pyramid with the leader on top and all the people on the bottom. The leader is picked from the bottom row and raised to the top. Power goes from the bottom to the top. You'll see in a minute that in some kinds of government, power goes the other way.

Democracy comes in two varieties: direct democracy and representative democracy.

When you choose someone to vote for you, you have ***representative democracy.*** The members of the House of Burgesses at Jamestown represented the colonists and made laws for them. Our Congress does the same thing for us today. We live in a representative democracy based on law. We are ruled by laws made by congressmen and congresswomen whom we elect.

Some New England towns have ***direct democracy.*** All the people in the town get together at a town meeting and vote directly on important issues. No one votes for them.

It is possible, with modern technology, that we will have more direct democracy in the future. We may be able to vote directly for certain laws through our TV sets. However, we will always need representatives who can take the time needed to make decisions on complicated issues.

Autocracy is the opposite of democracy. It is government by a single authority with unlimited power. In autocracies, power starts at the top of the pyramid.

A ***dictator*** is an autocrat. If a dictator doesn't like you, he can have you killed without consulting anyone or giving you a trial. Autocrats don't have to be bad. There have been a few good ones in history—but only a few. If a dictator is terrible, the people are stuck with him (or her). They have no power—except, sometimes, brute force. In the old days, kings and queens were autocrats. Today they usually share power with a parliament.

Alexis de Tocqueville was a Frenchman who came to the United States in the 19th century. He wrote a book called Democracy in America *that is still one of the best descriptions of our system ever published.*

A ***theocracy*** is...if you don't remember that, go back and read the last chapter. Just kidding, you don't have to do that. A theocracy (remember?) is government by a church in the name of God.

An ***aristocracy*** is government by a small group of privileged people. England was an aristocracy in the 17th century. The king and the landowning aristocrats (who controlled Parliament) ruled.

Hold on. This is tough, and may be a bit boring, but it is important.

Democracy, autocracy, theocracy, and aristocracy are all words that tell you about political power and who has it.

Here are two words that tell about economic or money power: ***communism*** and ***capitalism.***

In a communist country the state owns most of the land and property and shares them with its citizens. (Remember, the word *state* sometimes means "nation." That can be confusing, but that's the way it is.) People don't own their own homes or businesses. They work for the state. Jamestown tried a kind of communist system, and so did Plymouth. Both colonies found that people work harder when they know they can own land or a business. (In the 20th century, Russia and most of Eastern Europe tried and rejected communism.) People often need to be forced to be communists.

In a capitalist country you can own capital: capital is money and property. The United States is a nation that practices capitalism. In a capitalist country goods are distributed through a free market. Capitalism has disadvantages, too: wealth often piles up for a few people, while others don't have enough.

Whew, that's a lot to absorb! Now let's get back to history, which is much more exciting.

16 Of Towns and Schools and Sermons

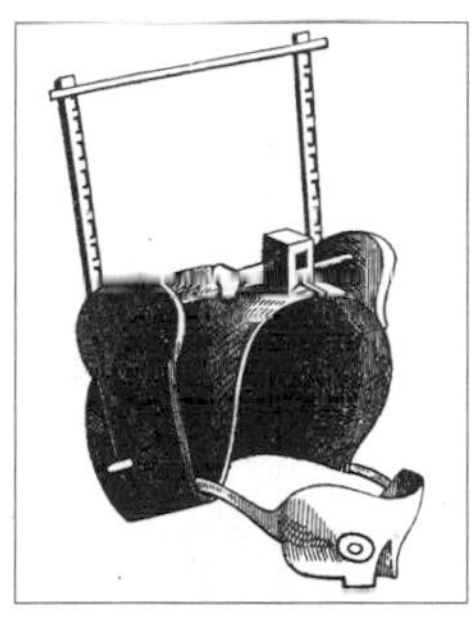

Women who nagged or talked too much—scolds—were considered a curse in the 17th century and could be made to wear a *scold's bridle*.

At first the New England settlers built their homes behind high fences called stockades. They were fearful of the unknown—of Indians and animals.

Soon they began spreading out, beyond the fences, into small towns with names like Greenfield, Springfield, and Longmeadow. The names described the land. Many of those early settlements were just a row of houses strung alongside abandoned Indian fields that the English settlers found and took. They lived with Indians as neighbors, although their animals sometimes made that difficult.

The Native Americans hunted animals; they had no horses, cows, sheep, or hogs. The Indians soon discovered that those English grazing animals could destroy their cornfields. In 1653 the people of the town of New Haven agreed to work for 60 days to build fences around fields planted by neighboring Indians. New England's courts had been ordering colonists to pay the Indians for damage done to their fields by wandering animals.

As the colonists began to prosper, they built towns in America that were something like the villages they left behind in Europe. They were compact, easy to defend, and friendly. Castles and manor houses dominated European towns; in New England's villages it was the meetinghouse that stood out. The meetinghouse was used as a church, a town hall, and a social center. It was usually placed at one end of a big field that was called a common, because everyone used it in common. Sometimes, when there were sheep to chew the field's grass and keep it short and green, the common was called a green. Houses were built

People in New England villages were usually friendly and neighborly to each other. They had to be. A family needed the neighbors' help to clear rocks out of a field or raise a barn roof. There was one cowman who looked after everybody's cows. But if a stranger came hanging around with no invitation from a local family, he was chased out of town.

Chairs were rare and costly in the 17th century. This one belonged to Governor Endicott.

The Puritans had so many rules and laws that they were continually being broken. Another popular punishment, especially for scolding women, was the ducking stool. The sinner was tied on and lowered into the stream or village pond.

Puritans liked to give their children names that were reminders of goodness and holiness. Some we still find occasionally, like Constance, Faith, or Hope; and some seem strange—Joy-from-Above, Kill-sin, Fear, Patience, Wrestling-with-the-Devil.

around the green. The houses nearest the meetinghouse belonged to the most important people in town: the minister and the church leaders.

Many villages had a stream. The tumbling water of the stream turned a big wheel, and that provided power for the mills where wood was sawed and wheat ground into flour.

As the town grew other buildings were added: a general store, a blacksmith's shop, a furniture maker's shop, a candlemaker's.

If the town was large enough, there might be an inn. Almost always there was a school.

The Puritans cared about schooling. By 1636 they had founded Harvard College. It was amazing that they had a college so soon after they arrived, although Harvard did get off to a rocky start. The first teacher beat his students, fed them spoiled meat, and ran off with college money.

Then they got a college president, Henry Dunster. He was so good that students began coming to study with him from Virginia and Bermuda and even from England

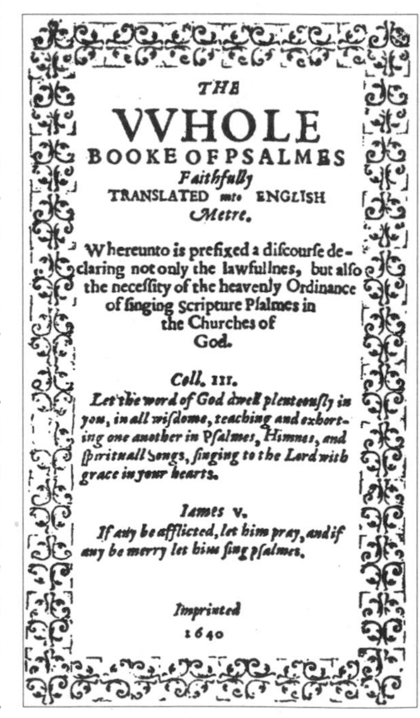

THE
VVHOLE
BOOKE OF PSALMES
Faithfully
TRANSLATED into ENGLISH
Metre.

Whereunto is prefixed a discourse declaring not only the lawfullnes, but also the necessity of the heavenly Ordinance of singing Scripture Psalmes in the Churches of God.

Coll. III.
Let the word of God dwell plenteously in you, in all wisdome, teaching and exhorting one another in Psalmes, Himnes, and spirituall Songs, singing to the Lord with grace in your hearts.

Iames v.
If any be afflicted, let him pray, and if any be merry let him sing psalmes.

Imprinted
1640

The *Whole Booke of Psalms*, also known as the Bay Psalm Book, was the first book ever printed in the English colonies.

A footwarmer for churchgoers.

itself. Of course, they were all Puritans.

Because of their religion, Puritans weren't allowed to attend college in England. That was one reason it was so important to have Harvard succeed. To do that it had to have a supply of students. So, in 1642, the Massachusetts Bay Colony passed a law saying that parents must teach their children to read.

The Puritans wanted everyone to be able to read the Bible, even those who weren't going to Harvard. So the next thing they did was pass a law that said:

> *It is therefore ordered, that every township in this jurisdiction, after the Lord has increased its number to 50 householders, shall then forthwith appoint one within their town to teach all such children as shall resort to him to write and read, whose wages shall be paid either by the parents or masters of such children, or by the inhabitants in general.*

In plain English, that means that every town with 50 or more people must have a schoolteacher.

Do you see something unusual in that law? Read that bit at the end, "shall be paid...by the inhabitants in general." Do you know what that means? It means that everyone in the town has to pay for the education of the children. Not just the parents. That is what public education is all about. It guarantees that every child, not just those with wealthy parents, can go to school. In America, it all began with that school law in 1647.

Oh, I know what you're thinking. Why did they have to go and do it? Who needs school anyway? But you don't really mean it. It isn't fun to be ignorant.

In the 17th century much teaching was done by parents, or in church, or, if you were an apprentice, by your master. But the Puritans could see that sometimes that wasn't enough. Some parents just weren't good teachers. The Puritans thought it important that everyone read the Bible. In Boston and the larger towns some children were actually taught to read the Bible in its

Blowing Thy Nose

Many little Puritan boys and girls had to study a book called *The School of Good Manners*. It reminded them to "stand not wriggling with thy body hither and thither, but steady and upright," or that "when thou blowest thy nose, let thy handkerchief be used." Naughty children were whipped with a birch stick or cane. "Spare the rod and spoil the child" was a firm belief even of kind parents and teachers.

Harvard College (above) had a very English class consciousness for many years. Until 1769 the roster of students was not listed in alphabetical order, but according to social status. That meant that if you were from an important family, you were listed ahead of somebody of lowly rank.

NEW ENGLAND VILLAGE
Chandler
Mill
Inn
Potter
Minister's House
Meetinghouse
Shoe Shop
Furniture maker
Stocks
Common
Well
Covered Bridge
Cooper
School
Blacksmith

Several generations learned their ABCs from the *New England Primer* (right), which used rhymes to help children remember letters. This might be all you got if you were a girl—very few had as much schooling as boys.

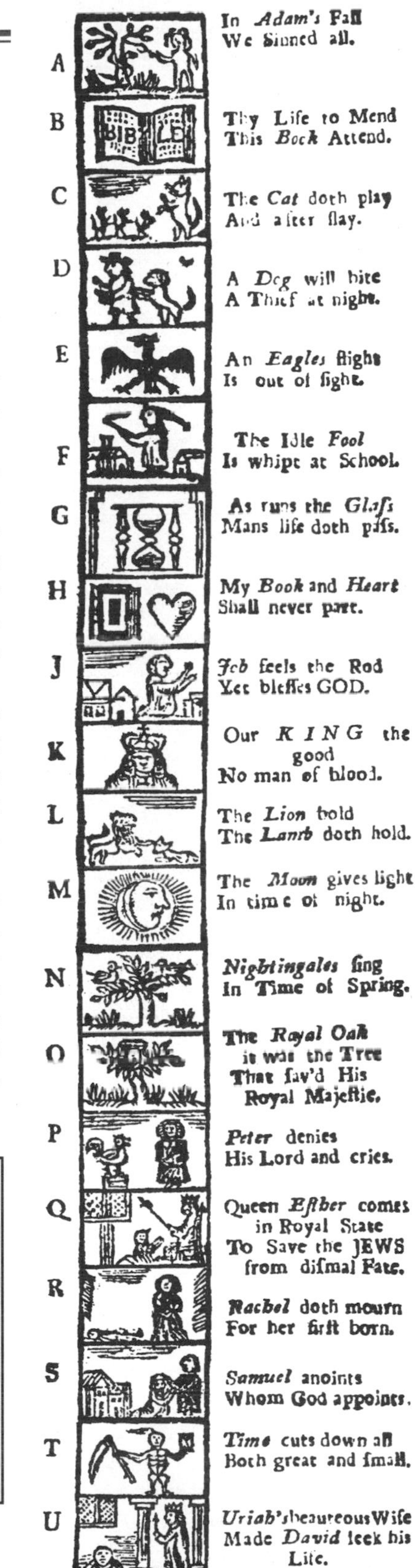

A In *Adam's* Fall
We Sinned all.

B Thy Life to Mend
This *Book* Attend.

C The *Cat* doth play
And after slay.

D A *Dog* will bite
A Thief at night.

E An *Eagles* flight
Is out of sight.

F The Idle *Fool*
Is whipt at School.

G As runs the *Glaſs*
Mans life doth paſs.

H My *Book* and *Heart*
Shall never part.

J *Job* feels the Rod
Yet bleſſes GOD.

K Our *KING* the good
No man of blood.

L The *Lion* bold
The *Lamb* doth hold.

M The *Moon* gives light
In time of night.

N *Nightingales* ſing
In Time of Spring.

O The *Royal Oak*
it was the Tree
That ſav'd His
Royal Majeſtie.

P *Peter* denies
His Lord and cries.

Q Queen *Eſther* comes
in Royal State
To Save the JEWS
from diſmal Fate.

R *Rachel* doth mourn
For her firſt born.

S *Samuel* anoints
Whom God appoints.

T *Time* cuts down all
Both great and ſmall.

U *Uriah's* beauteous Wife
Made *David* ſeek his
Life.

W *Whales* in the Sea

original languages. So little Puritan boys and girls of six and seven learned to read Latin and Greek, and a few learned Hebrew, too. That sounds hard, and it was, but learning languages is good training for the mind. Many of this nation's greatest thinkers came from Puritan stock.

Try and take yourself back to Puritan times, and see what you think of Sunday churchgoing. Those Puritan ministers gave sermons that lasted for hours and hours. Sometimes there was an intermission for lunch, and then everyone went back to hear more. There was no heat in the meetinghouse, and New England can get very cold. People brought warming boxes with hot coals in them to keep their feet from freezing. Sometimes they brought their dogs to church for the same reason.

A church official held a tickling rod to wake up anyone who looked as if he might be falling asleep. The dog whipper took out dogs who barked. If you were a troublemaker and wiggled and made noise you could get locked up in the town stocks. You'd have to sit there with your hands and feet stuck into a wooden contraption and everyone would make fun of you.

We know you wouldn't like that kind of life, but maybe things weren't so bad for the Puritan boys and girls. Maybe some of them even looked forward to the sermons. Remember, in Puritan Massachusetts there were no movies and no TVs. At first, there were no newspapers, no magazines, and only a few books. The Puritans were intelligent people who could read and think well. Maybe that will help you understand why everyone tried to listen to the weekly sermon and why Puritans sometimes spent all week talking about it.

Feeling Blue

Rules banning work, trade, and playing on Sundays—the Sabbath—are still called *blue laws*, because the Puritans wrote the laws in books bound in blue paper. You could be fined or punished for doing these things on Sunday: running, cooking, making a bed, or shaving. A man was whipped for saying that the minister's sermon was boring. Another was put in the stocks after kissing his wife on his return home from three years at sea. And celebrating Christmas was forbidden. It was "popish"—something that Roman Catholics did. Most Puritans worked on Christmas—unless, of course, it happened to fall on a Sunday.

17 Roger Williams

A Key into the
LANGUAGE
OF
AMERICA:
OR,
An help to the *Language* of the *Natives*
in that part of AMERICA, called
NEW-ENGLAND.
Together, with briefe *Observations* of the Customes, Manners and Worships, &c. of the
aforesaid *Natives*, in Peace and Warre,
in Life and Death.
On all which are added Spirituall *Observations*,
Generall and Particular by the *Authour*, of
chiefe and speciall use (upon all occasions,) to
all the *English* Inhabiting those parts;
yet pleasant and profitable to
the view of all men:
BY ROGER WILLIAMS
of *Providence* in *New-England.*

The first words Williams translated in his guide to the Narraganset language were *I love you.*

The Puritans, who were victims of intolerance in England, were not tolerant themselves. Although they preached the Golden Rule—do unto others as you would have them do unto you—they never understood that they were breaking that rule. Roger Williams did.

Williams was a Puritan minister who came to Massachusetts seeking a "pure" religious community. Like the other Puritans, he was a serious Christian. Like the others, he disapproved of Catholic and Quaker ideas. Like the others, he thought the Indian religions were pagan.

Catholics and Quakers are also Christians. They just interpret Christianity in a different way from the Puritans.

But that's where "like the others" stops.

He didn't believe in forcing anyone to believe as he did. He believed that killing or punishing in the name of Christianity was sinful. He thought that church members—not general taxes—should pay the bills at each church. He respected the beliefs of others. Those were strange ideas in 17th-century Massachusetts.

The Puritans didn't know what to do with Roger Williams. He was a Puritan, he was brilliant, he was a minister, and he was so nice that even his opponents had a hard time disliking him.

But what an ungrateful young man he seemed! The Puritans had offered him good jobs, as teacher and minister, and he thanked them by criticizing their practices.

Governor Winthrop was shocked. So was John Cotton, the minister who took a job that Williams refused. They were especially shocked when Williams wrote a book saying it was wrong to persecute people for their beliefs. Williams called his book *The Bloody Tenet.* (The blood was from those killed because of their religious ideas.) John Cotton wrote his own book. He called it *The Bloody Tenet Washed and Made White.* Of course that

To the Puritans, ***pagan*** religions were not "real" religions, like Christianity.

didn't end it. Williams's next book was *The Bloody Tenet Made Yet More Bloody by Mr. Cotton's Endeavor to Wash It White in the Blood of the Lamb.*

When Roger Williams started preaching that land shouldn't be taken from the Indians—that the king had no right to charter land that didn't belong to him—that was too much. The officers of the Massachusetts Bay Colony made arrangements to ship Williams back to England. They sent armed men to put him on a boat.

Roger Williams's wife heard the verdict of the court—that he was to be arrested and banished—and began to cry. Williams told her, "Fifty good men did what they thought was just." Roger Williams didn't hold grudges.

But he wasn't about to let himself get shipped back to England, and so he fled from Massachusetts. It was January 1636, he was sick, and the weather was freezing. Later, when he was an old man, Roger Williams would still remember that terrible winter. He was always thankful to the Narraganset Indians, who helped him survive the cold. He learned to love them as they loved him.

He bought land from the Indians and started a colony called Providence. It became the capital of Rhode Island and soon attracted many of those who were not wanted elsewhere. Someone described Providence as the place where "all the cranks of New England" go.

There were plenty of cranks in Providence. But there were also many people who were searching for what Roger Williams called "freedom of conscience."

In 1763, thanks to the atmosphere of tolerance that Roger Williams fostered, Rhode Island became the home of the first permanent Jewish temple in America, the Touro Synagogue.

When Roger Williams said freedom of conscience, he meant the freedom of each person to follow his own mind and heart and choose his own religion. That was to become an important right in America.

Roger Williams welcomed everyone who wished to come to Rhode Island, including Quakers and Catholics. And, while he continued to disagree with those religions, he never let that stop him from liking some of the people who practiced them. Jews, who were often persecuted elsewhere, were welcome in Rhode Island. Atheists were welcome, too.

Williams believed that state governments should not have any connection with a church. We call that separation

John Cotton (above) and most Puritans thought Williams's ideas of toleration were wrong. Maybe they thought he was crazy, too.

A ***tenet*** (TEN-it) is a basic idea, a fundamental concept. ***Atheists*** believe there is no God.

The poet means "meat" when he says ***flesh***. When he says ***they part to friends***, he means "they share with friends." He means "to lack humanity" when he says ***to want humanity***. Languages keep changing, and some English words were different in the 17th century.

of church and state. It was a very new idea at the time.

He knew that people could be forced to go to church, but that no one's mind could be forced to believe. "Forced worship stinks in God's nostrils," said Roger Williams.

He learned the language of the Narraganset Indians and wrote a book so that others could learn it, too. In it he included these rhymes about the Indians:

Sometimes God gives them fish or flesh,
Yet they're content without.
And what comes in, they part to friends
And strangers round about.

If nature's sons both wild and tame
Humane and courteous be,
How ill becomes its sons of God
To want humanity!

When the great Narraganset chief Canonicus (kuh-NON-ih-kuss) was dying, he called for Roger Williams to be with him. White men had destroyed the Indian chief's kingdom, and he hated most of them. But Williams and Canonicus had something in common. Each was able to judge people by their character, not by their skin color or religion. They loved and respected each other.

Edmund S. Morgan, who wrote a book about Roger Williams, said "We may praise him...for his defense of religious liberty and the separation of church and state. He deserves the tribute...but it falls short of the man. His greatness was simpler. He dared to think."

Roger Williams

Church and State

The Puritans forced some Indians to become Christians. Roger Williams wrote a letter to the Massachusetts governor. "Are not the English of this land generally a persecuted people from their native land?" he asked. How could those who had been persecuted persecute others, he wondered? He said that the Indians should "not be forced from their religions."

Roger Williams didn't think that anyone should be compelled to follow a religion. Besides, he knew it never works to try that. You can make people do things, but you can't make them believe what they don't want to believe.

Williams said that it was "against the testimony of Christ Jesus for the civil state to impose [force] upon the souls of the people a religion."

Most Puritans didn't agree with Roger Williams. They thought it was the job of the government leaders to tell people what to believe.

But Roger Williams's ideas won out. They helped bring about the separation of church and state that is one of the most important of all of America's governing ideas. In Europe and the rest of the world, millions of people have died in wars over religion, but that has not happened in this country.

Roger Williams, a devout Puritan, wrote, "Jesus never called for the sword of steel to help the sword of spirit."

18 "Woman, Hold Your Tongue"

Anne Hutchinson's clothes were plainer than these. But still she had to wear underskirts and petticoats. Imagine wash day, and no washing machine.

Anne Hutchinson was another troublemaker. At least that is what some Puritans thought. Here was a woman with 14 children who was interpreting the Bible. No one objected to that, until she began to question some of the ministers' beliefs. Soon she was trying to reach everyone with her ideas about God. Governor Winthrop was outraged. Didn't she have enough to do, with all those babies to feed?

What was worse, in Winthrop's view, was that people were listening to her. Even men were listening. There was a reason: Anne Hutchinson had a fine mind, and she loved God. Besides, what she was saying made sense. Winthrop admitted that she was "a woman with a ready wit and bold spirit." Before long, Massachusetts was split between people who believed what Anne Hutchinson said and those who believed the ministers. She claimed God was guiding her; the ministers said they were doing God's work on earth.

Finally, the Puritans held a trial. You can read the court records for yourself. You may agree that Mrs. Hutchinson was smarter than her accusers, but that didn't help her a bit. She was kicked out of Massachusetts and out of the Puritan church, too. She moved to Rhode Island. Later Anne Hutchinson moved to New York and was killed by Indians. Governor Winthrop saw it as the judgment of God.

In those days, women, like children, were expected to be seen but not heard. They belonged to their husbands. The word for them was *chattel*. That means a piece of property. A husband could sell his wife's labor and keep the wages. If she ran away, she was accused of stealing herself and her clothing. Her husband even owned her clothes.

From Me to Thee

Puritans and Quakers used the words *thou* and *thee* instead of "you," and *thy* instead of "your." They are old-fashioned words that people used when talking to a child, a close friend or family member, a servant—or God. What the Puritans were saying with those words was that we are all close; we are all brothers and sisters.

No More Diapers

You might think that Anne Hutchinson had a terrible lot of diapers to change with all those children. But as soon as a baby could sit up its tight swaddling bands were taken off. In warm rooms and climates, babies didn't wear anything—though sometimes a mother wrapped a "baby clout" (probably just a piece of linen) around a baby's bottom when it sat on somebody's lap. In colder weather, babies and little kids mostly just wore a shirt—with nothing underneath. When toddlers had to go to the bathroom, their mothers plunked them down outside. Not very hygienic, maybe, but simple.

19 Statues on the Common

This Quaker was arrested for preaching, driven out of town, and beaten with a cat-o'-nine-tails.

Anne Hutchinson wasn't the only strong woman to trouble John Winthrop and the Massachusetts Bay Puritans. Her best friend was a problem, too.

Mary Dyer was Puritan and pious. Winthrop called her a "very proper and fair woman." That was when he first knew her. But Dyer followed the ways of Anne Hutchinson, and when Anne Hutchinson was cast out of Massachusetts, Mary Dyer and her husband, William, and other believers went with her.

Later, Mary Dyer took a trip back to England and found other truths for herself. She became a member of the Society of Friends, the people who were known as Quakers. Quakers call their church services meetings. In a Quaker meeting everyone is equal, anyone may speak out, and there are no ministers. Now, in the 17th century equality was not fashionable. Besides, Quakers refuse to swear oaths of allegiance to anyone but God. But oaths of loyalty to king and country were expected in England and everywhere in the 17th century.

You need to understand that the church and the government were all part of one package in the Old World. It was the church that gave the king his right to govern. It was called the divine right of kings. It was the government that gave the church support and lands. That is the way it had always been. It seemed as if people like the Quakers

At Quaker meetings, the congregation sits and meditates in silence. Sometimes a member feels that God is communicating with him or her directly. The Friend might start talking aloud about this "inward light," or might shake and tremble—which was how the Quakers got their name.

At Puritan church services women sat at the back or upstairs. At a Quaker meeting (right) everyone sat together.

wanted to mess things up. The Quakers believed in toleration, and they believed each person could think for himself. What happens if you let people think for themselves? Why, the next step might be for them to say that the king's church and the king's priests weren't needed. And Quakers did say something like that when they sat in their meetings without ministers.

So maybe you can see why Quakers were hated and persecuted by the authorities in England. They weren't liked any better in the colonies. The magistrates of the Massachusetts Bay Colony passed harsh laws to keep them away, but that didn't stop them. Some Quakers seemed determined to be martyrs, and Mary Dyer was one of them. She came to Boston and was sent away. She came back. This time she was tried, with two Quaker men, and all three were led to the gallows.

A hanging, like that of Mary Dyer, was a public spectacle in the 17th century. People took time off work to watch. They cheered, jeered, and partied.

The men were hanged, but at the last minute Dyer was put on a horse and sent off to Rhode Island. She came back again. Now what do you do with a woman who cares so much about her religion that she will risk death to preach its message? You'd think the Puritans would have understood that devotion. Maybe they did, and that's what scared them.

The Puritans tried Mary Dyer again. This time they offered her her life if she would leave Massachusetts forever. She refused. Mary Dyer was hanged, on June 1, 1660, on the Boston Common in front of where the State House stands today. Her death was too much for some Puritans. In 1661 the law was changed. Today, statues of Mary Dyer and Anne Hutchinson can be seen on the Common.

But most Puritans thought they had done everything they could to be fair to Mary Dyer. Remember, it was a different world then, a world just leaving the Middle Ages. A few people in Europe were beginning to question the old ideas. But those questions traveled slowly across the ocean. When people started talking of toleration, the Puritans "could hardly understand what was happening in the world," writes Perry Miller, a historian of the Puritans, "they could not for a long time be persuaded that they had any reason to be ashamed of their record of so many Quakers whipped, blasphemers punished by the amputation of ears, [dissenters] exiled...or witches executed. [According to the beliefs] in Europe at the time the Puritans had left, these were achievements to which any government could point with pride."

A ***martyr*** (MAR-tur) is someone who would rather die than give up his belief.
A ***gallows*** was the two standing poles and crosspiece from which people were hanged. Sometimes bodies were left on the gallows for a while, as a warning to others.

20 Of Witches and Dinosaurs

Everybody believed in witches. James I wrote a book that said they should be put to death.

How many Puritans do you know? Don't think too hard. The answer is zero. Puritans are like dinosaurs: they are extinct.

I'll tell you what happened to them in just a minute, but don't worry, things worked out well. Many of their descendants turned into New England Yankees. Others are spread all across the nation. Many go to the Congregational church. In some ways, however, we are all descendants of the Puritans.

Now you may be shaking your head if you live in California, or Colorado, or West Virginia. If your parents came from Mexico or China, you're probably saying, "No way am I a Puritan!" But it's true. If you are an American, you are a descendant of the Puritans—at least a little bit—because many American laws and ideas come from Puritan laws and ideas, and they are some of the best we have.

You see, the Puritans hoped to build a place on earth where people could live as the Bible says they should, a place where people would be truly good. Governor Winthrop called Puritan Massachusetts "a city upon a hill." He expected it to stand tall as a symbol to the rest of the world. "The eyes of all people are upon us," he said.

The Puritans came to the New World to try and build a godly community where they could live close to perfect lives. No human beings have ever been able to do that, but the Puritans tried.

To make their community pure, the Puritans ex-

Matthew Hopkins, England's Witch Finder General.

That phrase, "city upon a hill," is one to put in your head and remember. You'll hear it again and again when you read American history.

Do you know of other societies or times in history where neighbors spy on neighbors—sometimes, even, where children spy on their parents? Do you think that is likely to be a good thing for the people involved?

pected everyone to act like a spy and report any neighbor who did or said anything wrong. Can you see how that might make you a little uncomfortable around your neighbors? Maybe that helped cause the Puritans' downfall. Or perhaps it was their seriousness that did them in, or witches (we'll get to them soon), or their self-righteousness.

Self-righteous people believe they know the truth. They think that anyone who doesn't agree with them is wrong. They are apt to judge other people. Sometimes the Puritans were like that. They spent time judging their neighbors.

They read the Bible every day, but they didn't always pay attention to the passage in the New Testament that asks: "Why do you observe the speck of sawdust in your brother's eye and never notice the log in your own?" (What does that mean?)

A witch and her imps. An imp could be a monster, but more often it was an ordinary animal—even a fish—with horns or wings.

Benjamin Franklin, one of the wisest of all Americans, wrote in his *Autobiography* a warning against using words like "certainly, undoubtedly, or any others that give the air of positiveness to an opinion." Instead, said Franklin, we should say, "it appears to me, or I should think it so or so...if I am not mistaken."

The Puritans were very sure of themselves. They believed that God saves only a very few people and that the rest go to a terrible hell filled with fires. They thought that God decided when a child was born if he was saved or not. They used the word "elected" instead of saved, and they thought they were God's elect. Because of that, they thought they should act like God's elect and lead good lives. And that's why we can be happy they were among the first colonists. They really tried their best to be good. They worked hard, they believed in learning, and they did what they thought was right.

Cotton Mather

wrote: "And we have now with Horror seen the Discovery of such a Witchcraft! An Army of Devils is horribly broke in upon...our English Settlements: and the Houses of the Good People there are fill'd with the doleful Shrieks of their Children and Servants."

Witch Advice

IF THE CREAM WON'T TURN TO butter, a witch must be in the churn. Take a horseshoe, heat it red hot, and throw it into the churn. That will set the witch flying.

NAIL A HORSESHOE OVER YOUR door, and witches will stay away.

TO TELL IF SOMEONE IS A WITCH: tie the suspect's hands and feet. Toss him (or her) in the water. If he sinks he is innocent. If he floats he is a witch. (Which would you rather be?)

No one made pictures of the Salem witch trials while they were happening, so we can only imagine what they were like from what was written down. This artist imagined the trial of George Jacobs, the old man kneeling on the right. (Yes, men could be witches, too.)

But the Puritans had a big problem. They were human and they made mistakes. Their ministers didn't help. The ministers said that people were naturally sinful, but if they sinned they would go to that terrible hell. That kept everyone under constant pressure to be close to perfect. No one could relax. And that may be what caused the nightmare of the witches.

The whole world believed in witches—there was nothing new about that. People thought that if you wanted to make a bargain with the devil you could do it, and then torment people and fly through the air on a broomstick, or become invisible and squeeze through keyholes. Everyone *knew* that witches could create thunder, sink ships, kill sheep, and make tables and chairs rattle.

Back in the days of Christopher Columbus, Pope Innocent VIII had issued orders that witches were to be burnt. So in Switzerland, during three months in 1515, more than 500 witches went to the stake. Soon fires were lit in France and Germany and all over Europe. England's King James wrote a book about witches, but he wasn't alone. London's publishers were busy with the subject. In the 17th century, Parliament

Judge Sewall sentenced several witches to death. Later he apologized.

appointed Matthew Hopkins as witch finder. And Hopkins found a lot of witches. Anyone with a pimple or a wart or a mole was a suspect. Those were said to be devil's marks. It was amazing—after some torture, many people confessed to knowing the devil.

In the colonies some men and women were hanged or drowned for witchery, but what happened in Salem, Massachusetts, was different from the usual story.

It all began with some little girls and their servant, Tituba. Now Tituba was a poor woman from the West Indies who told stories of the devil and witches and voodoo. The girls were nine-year-old Elizabeth Parris, eleven-year-old Abigail Williams, and their friends. Tituba's stories must have been scary, especially around the fire at night. But when Tituba taught the girls to bark like dogs and mew like cats and grunt like hogs, that might have been fun. Although Elizabeth's father, the Reverend Samuel Parris, didn't think it was fun at all. When he saw the children grunting and mewing and sometimes acting as if they were having fits or spasms, he remembered reading books from England about spells laid on people by witches. He became alarmed.

Four witches are hanged in this 1655 picture. That is the hangman on the ladder. The man on the right is the witch finder, taking his money.

And then, to every one's astonishment, on Sunday the girls spoke out during church meeting and said silly things. "There is a yellow bird on the minister's head," cried Anne Putnam. No one would interrupt a church service except the devil! So when the girls said Sarah Good, Sarah Osburn, and Tituba were bewitching them, everyone believed them. The two Sarahs denied the charge. But they were old and poor and no one listened to them, especially after Tituba said that she did indeed fly through the air on a broomstick.

The little girls might have been pretending when they started, but

Evil Explained

It wasn't only people from European cultures who feared witches or killed people suspected of witchcraft. Some Native Americans did the same thing. It happened among people who—like the Puritans—were very religious, and wanted an answer to an overwhelming question.

If God is good, how do you explain evil? That is the big question that has concerned people from the beginning of time. It is a central question of great literature and philosophy and religions in all cultures.

There is an easy answer: witches. Suppose you can't explain a sudden earthquake that destroys your town. Suppose you can't explain a disease that kills the young and sweet and promising. Suppose you can't explain a blight that settles on the corn crop and destroys your winter food supply.

Witchcraft is the easy answer. Like the Puritans, the Indians of the pueblo explained the unexplainable with witches. In times of crisis they accused some among themselves of witchcraft, and then they killed the accused. Sometimes one or two witches were enough. Sometimes, as at Salem, it took as many as 20 to exhaust the people's fear.

The Reverend Cotton Mather was a minister and a member of one of Puritan New England's most important families. He was a scholar, and one of the first Americans to promote the smallpox vaccine when most thought it dangerous. Yet he believed firmly in witches and encouraged the hysteria in Salem.

soon they were telling of torture and witchery, and perhaps they convinced themselves. (Or maybe, now, they were afraid to tell the truth.) Their stories grew longer and their screeches louder. They accused one person after another of putting spells on them. People came from all over Massachusetts to watch them squeal and grunt. What was to be done? Suddenly, anything that went wrong in that little town was the fault of a witch. Salem was mad with witch fever. Five-year-old Dorcas Good was taken to jail and chained to her mother when the girls said she was tormenting them. Then other people began talking of witches and pointing at their neighbors. A court was called to hear the evidence.

The judges were scared, like everyone else. The leaders of the community, who might have done some thinking, didn't. More than 100 people were tried as witches; 20 people and two dogs were put to death. Then the Reverend Hale's wife was accused. But there was no one in Massachusetts more beloved and godly than she! Could it be that the girls were wrong? Everyone had believed them; no one had believed the victims. Were those people who had been killed innocent? Yes, an awful tragedy had occurred.

The witch trials were a shameful chapter in American history, although when one of the judges, Samuel Sewall, realized the wrong that had been done, he publicly apologized. The court cleared the names of those who had died. That does make this story better than some. It was a time of witch fear. There had been many witch hangings and burnings in Europe and America; never before was there a public apology. But something devilish did die in that little town of Salem. It was the belief in witchcraft. Most people had learned a sad lesson.

Some Hard Questions

The Puritans have a bad image today. People think of them as narrow bigots. That's not quite fair. The Puritans were sophisticated, sincere believers who thought they had found the truth. They were people who wanted to do right in what they saw as an evil world. The Puritans had left Europe because they didn't want their children tempted by worldly society. But that society followed them. There was to be no escape from the real world.

The problem—for people who believed they knew the ultimate truth—was how to react when their "truths" seemed wrong.

The witchcraft crisis was a problem. So was the hanging of Mary Dyer, the expulsion of Roger Williams, and the persecution of countless Puritans who were thought to have "sinned." Their children and grandchildren began asking questions for which there were no good answers. The old religion was not to survive in the new land.

21 Connecticut, New Hampshire, and Maine

The name Connecticut comes from a Mohican word, *quinnitukqut*, meaning "at the long tidal river."

The New Haven colonists made their own rules. One said that every male had to have his hair "cut round."

No one is ever all bad or all good. The witch trials may have been a low point for the Puritans, but mostly these were good, strong, intelligent people. If they hadn't been, they never would have crossed the ocean; they would have stayed in England, as most English people did—even when they didn't like the way they were ruled.

Because the Puritans were independent thinkers they sometimes disagreed with each other. When they did, there was plenty of room in America. They could just start a new settlement.

Thomas Hooker was minister in a little town near Boston when, in 1636, he decided to move west; 100 of his followers went with him. He went to the beautiful valley of the Connecticut River. Today, no one is quite sure if he moved because of religious disagreements—or because the valley was fertile and farming was easier than in rocky Massachusetts. Perhaps it doesn't matter. He found good farming, and he was free to preach as he wished.

Hooker arrived in Connecticut without a charter, because the king hadn't given him permission to be there. He moved in anyway. He had no legal right to the land in the eyes of the English (and certainly none in the eyes of the Native Americans), so the Connecticut settlers sent a representative to England and got a charter. It allowed them to govern themselves, which was something English colonists always seemed to want to do.

Some people from Plymouth were already in Connecticut trading with the Indians. A few Dutch people were there too. Soon Puritans were coming from England and heading straight for Connecticut. Massachusetts was getting crowded and the land in Connecticut was

An artist's idea of the emigration of Rev. Hooker and his followers to Connecticut. Mrs. Hooker is in a litter—which was like a covered bed with poles to carry it by.

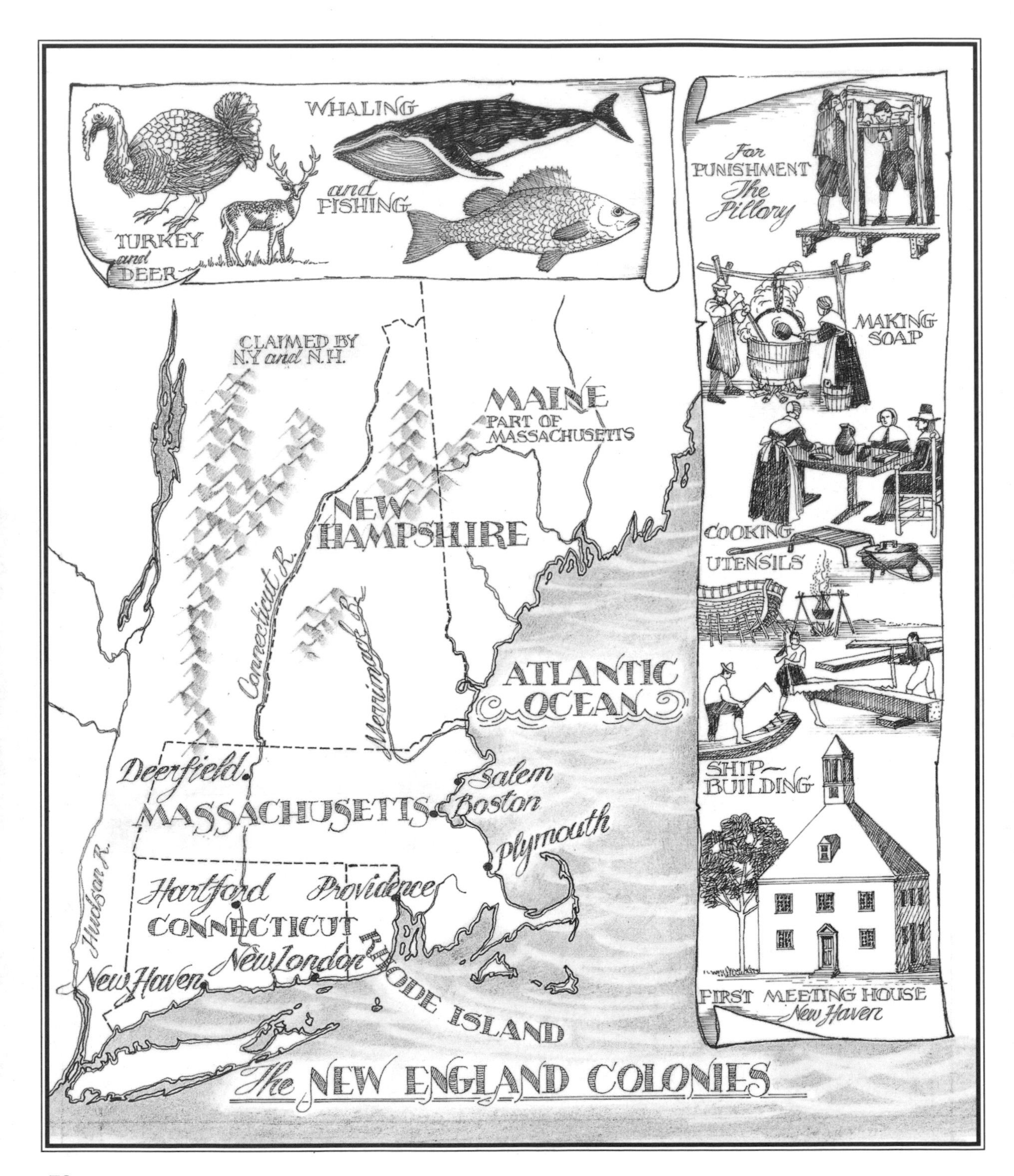
WHALING
and FISHING
TURKEY and DEER
For PUNISHMENT The Pillory
MAKING SOAP
COOKING UTENSILS
SHIP-BUILDING
FIRST MEETING HOUSE New Haven
CLAIMED BY N.Y. and N.H.
MAINE
PART OF MASSACHUSETTS
NEW HAMPSHIRE
Connecticut R.
Merrimack R.
ATLANTIC OCEAN
Deerfield
MASSACHUSETTS
Salem
Boston
Plymouth
Hudson R.
Hartford
Providence
CONNECTICUT
New London
New Haven
RHODE ISLAND
The NEW ENGLAND COLONIES

inviting. A group settled at New Haven in 1638 and another at New London in 1646. Each town had its own minister.

The New Haven colonists published a list of laws that said how you were to behave. One law said "Every male shall have his hair cut round." Another said, "Married persons must live together, or be imprisoned."

John Winthrop II, the son of the Massachusetts governor, became governor of the new settlements in Connecticut. He was kindly, much loved, and a friend to Roger Williams. When the Quakers were hanged in Boston, John Winthrop (who was called "the younger") pleaded for their lives "as on his bare knees."

While some people were going west to Connecticut, others were moving north, where there was a large piece of land the king had given to two friends: John Mason and Ferdinando Gorges. They divided that land: Gorges taking what became Maine, and Mason taking New Hampshire.

Since Mason and Gorges wanted settlers for their colonies, they did a sensible thing. They advertised for the settlers in England, and people came in response.

At first, New Hampshire was part of Massachusetts, with John Mason as its proprietor. Then it became a separate colony, and finally the King took it back and held it himself as a royal colony.

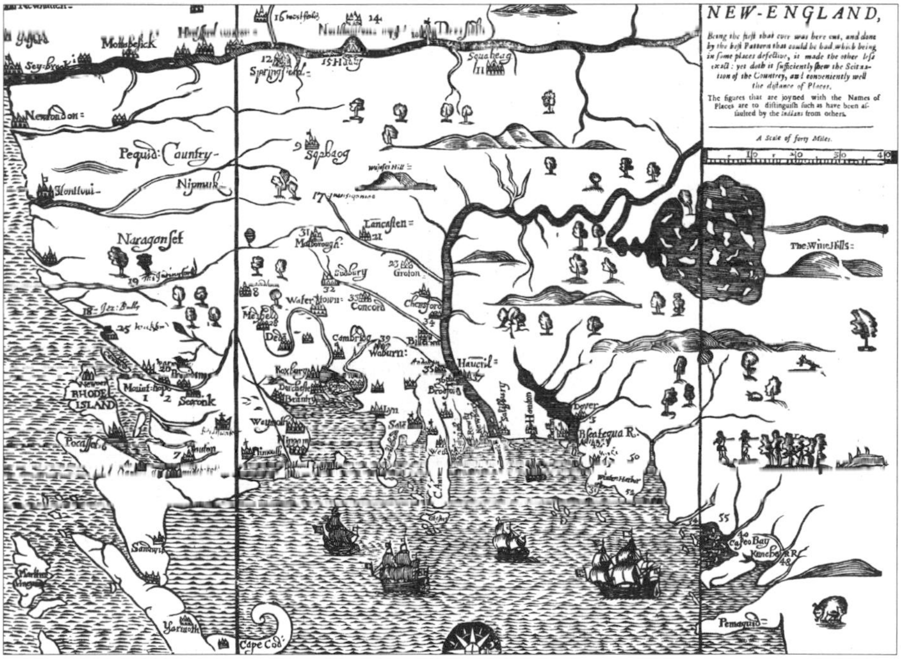

Maine was never a separate colony; it was part of Massachusetts until 1820, when it became the 23rd state.

As soon as colonies were established, they began competing for good land and good people. The more people who came, the more land that was needed.

The big losers in this contest were the Indians. Serious colonization could not take place until the Indians had been pushed off the land. And pushed off they would be.

On these pages are two artists' versions, old and modern, of New England and its coastline in colonial times. In the old map, the big blob of water with islands in it near New Hampshire's White Mountains ("White Hills") is Lake Winnepesaukee.

John Mason named New Hampshire after the county of Hampshire, which is on England's south coast.

Land Greed

It was the land that confused the Europeans—there was so much of it. They didn't know how to react. In Europe only the really wealthy—the aristocrats—could own land. Ordinary people didn't even dream of their own land. In America the land was so vast it would take the Europeans more than 200 years just to know how much there was. For a very long time, almost every free person who came to America could afford land.

Most Europeans couldn't quite believe their good fortune. They brought Europe's land greed with them. Each of them wanted to own land—often more than he needed. And almost as soon as the European-Americans got land, they wanted to change it, as Europe's land had been changed.

And so they looked at the beautiful forests and saw an enemy that needed conquering. And they conquered. They cut down trees, leveled hills, filled in swamps, and killed animals and birds. They didn't understand how to work with nature, as the Indians often did. They forced nature to conform to their ways and wants.

It turned out that the land wasn't endless, as they seemed to think at first. One day it would almost all be tamed. Then they would wish for some of those thick forests, some of those songbirds, some of the native animals. There would be few of them left.

The Indians didn't accept the European idea of landownership. Their religions taught them that the land and waters and animals belonged to God. They thought land could only be shared, not owned. So at first, when they signed treaties selling land, they thought they were selling the right to share it. They didn't expect to be ordered off the land.

Their beliefs told them to live in harmony with nature. Land was to be used by a tribe as a whole, not owned by individuals. It was a way of looking at the land that did away with most greed, but not all greed. Tribes often fought each other for control of the use of land.

Of course, if large numbers of people were to live on the land it had to be changed. Millions and millions of people can't live in a forest. Cities were needed for all the people who would come to live on this bountiful land. And so we built cities and suburbs and in the process often polluted and burned and destroyed. Did we have to do that? Can we have cities and also sheltering woodlands and clean rivers and abundant wildlife? Yes, but it isn't easy. We have to care about our environment. We have to respect the natural world.

Some say one answer is to put heavy industry on space platforms. Some say high-speed trains can be put underground. Some say ideas in science-fiction books can become fact and make the environment healthier.

Can you do anything to help? Of course you can. Do you throw trash around? Does your family worry about pollution? Have you ever helped clean up your neighborhood? Do you know what "conservation" means?

This View Towards Canaan and Salisbury in Connecticut *was made in 1789, but it conveys that feeling of the early settlers—that the land in America went on forever.*

22 King Philip's War

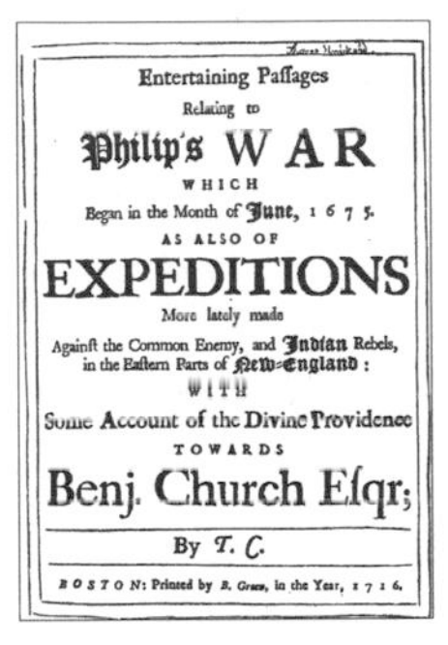
Entertaining Paſſages
Relating to
Philip's WAR
WHICH
Began in the Month of June, 1675.
AS ALSO OF
EXPEDITIONS
More lately made
Againſt the Common Enemy, and Indian Rebels, in the Eaſtern Parts of New-England:
WITH
Some Account of the Divine Providence
TOWARDS
Benj. Church Eſqr;
By T. C.
BOSTON: Printed by B. Green, in the Year, 1716.

The "entertaining" memoirs of Benjamin Church, who led colonial forces against King Philip.

Massasoit was a friend of the English colonists. The first New England settlers might not have survived without his help, and they knew it. Once, when he seemed near death, a group of settlers came from Plymouth with goose soup and a broth made from the root of the sassafras tree. Massasoit got better.

Massasoit's people, the Wampanoags, were hunters and fishermen and farmers whose lives turned with the cycle of the seasons. They were peaceful people and good neighbors. When some Pilgrims visited his village, Massasoit honored them by letting them spend the night on a plank bed with himself, his wife, and two of his chiefs.

A year after the Pilgrims arrived, Massasoit signed a treaty of peace with them. For more than 50 years, while he lived, there was peace in Massachusetts. But, even before he died, there were some—Indians and English—who saw trouble ahead. Mostly it was because there were so many Englishmen and women. At first there had been only a few of these newcomers, but soon they were pushing the natives off the land.

Massasoit's two sons were troubled. Their generation was different from that of their father. They were not awed by the English, as their father sometimes seemed. The two boys were Wamsutta and Metacom, but Massasoit had asked the General Court in Plymouth to give them English names. So they were named for ancient kings of Greece: Alexander and Philip.

When his father died, Wamsutta-Alexander became ruler. Some Englishmen feared him. They sent troops, dragged him to Plymouth, threatened him, and acted haughty and superior.

Many settlers survived King Philip's War only to die of starvation because the fields were trampled or never got planted. The Massachusetts Bay Colony almost went bankrupt, and 12 towns vanished.

As the Indians' land shrank, King Philip (above) told a friend, "I am resolved not to see the day when I have no country."

Alexander became ill and died on his way home. Metacom-Philip was now leader of his people. He believed the English had killed his brother, and he wanted revenge.

Besides, Metacom saw that the new people were destroying his land. (The English now called him King Philip. Some meant it respectfully, but others were mocking when they used the title.) And so Metacom began visiting other Indian leaders trying to convince them to join with him to fight the English and drive them from America. That wasn't easy. There was no history of Indian unity. The Indian peoples were as different from each other as Swedes are from Spaniards, or Chinese from Pakistanis. They were descended from different peoples who came in different waves of immigration over the Bering Strait.

Metacom wasn't ready when war began. As with many wars, it was really an accident that started things. A Christian Indian named John Sassamon was killed. Sassamon had been to Harvard and was a friend of the Plymouth colonists. Today, no one is sure who killed him, but three members of Metacom's tribe were executed for the murder. Metacom was furious. He attacked for revenge.

King Philip's War had begun. It was fought, off and on, for two years, 1675 and 1676, and it was horrible. If you have read about Indian wars—with scalpings, torched villages, tomahawks, and war whoops—you may have been reading about this war. Both sides were incredibly brutal.

Six hundred colonists lost their lives in King Philip's War; 3,000 Indians lost theirs. Fifty of ninety English villages were attacked; many were burned to the ground. The peaceful Narraganset Indians, who had nothing to do with the war, were massacred on their own land in Rhode Island because some settlers now feared all Indians. Many innocent whites

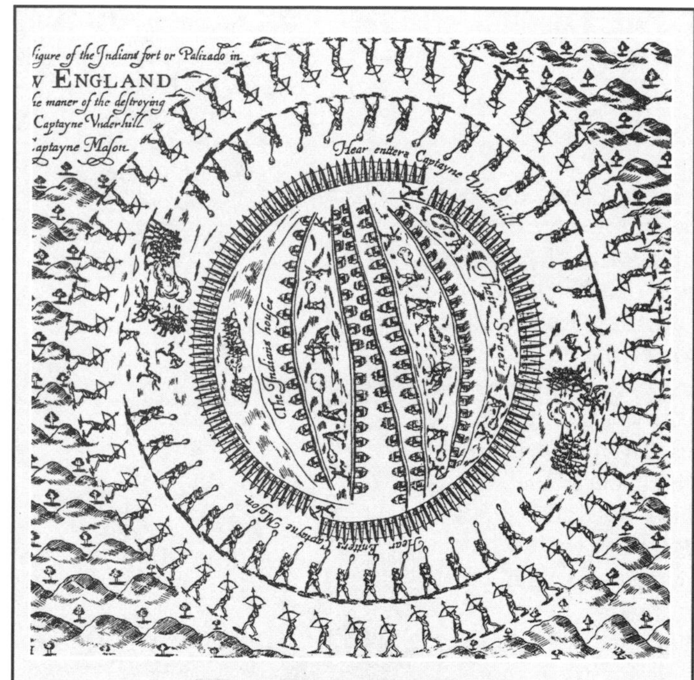

The Pequot War

The Pequots were Indians who lived in New England near Narragansett Bay. Settlers moved into their territory until the Pequots controlled less and less of it. The Indians got angry about this, and killed some settlers and traders. In 1636, the colonists retaliated (fought back) by destroying a Pequot village. The next year war broke out. Captain John Mason (who founded New Hampshire), and his allies from the Mohican and Narraganset tribes, attacked the Pequots' fort, near what is now West Mystic, Connecticut. Many Indians were burned alive. Those who didn't die were sold into slavery. The Pequots were almost wiped out. It was a taste of what was to come in King Philip's War.

An artist's rendering of KIng Philip's death in a swamp in Rhode Island.

were killed in Indian raids of revenge.

Indian disunity hurt their cause. Some tribes helped the English. In addition, Indian warriors weren't used to long wars. They knew how to attack and destroy in quick raids. When the war went on and on, many Indians got tired of it. They wanted to plant their crops and get back to normal activities. They deserted their leader. Finally Metacom was trapped in a swamp, where he was killed by an Indian who was loyal to the colonists. Metacom's head was chopped off and hung on the fort at Plymouth; there it stayed for 25 years. His wife, children, and other captured Indians were sold in the West Indies as slaves.

It was a pattern that was repeated over and over again until the Indians could fight no more.

Tomahawk comes from an Algonquian word for a war club, *tamahakan*.

This skeleton was found in a shallow grave in South Glastonbury, Connecticut. It is thought to be the remains of a Narraganset Indian shot in King Philip's War. A doctor who examined the skeleton in 1959 said that the man was a runner. He died of peritonitis, which developed as a result of his wound.

23 The Indians Win This One

The Spaniards felt it was their duty to convert Indians to Christianity. Their missions did double duty as military forts.

At the very time that Metacom was fighting to free his land from the English invaders, an Indian leader, a man named Popé (poe-PAY) was preparing for the same kind of fight. Popé lived far away, thousands of miles across the continent, in a place the Spaniards called New Mexico. He was a religious leader and a great medicine man, and he would win his battle. He would drive the enemy from his country.

It is 1676 and revolution is brewing in the valley of the river that the Indians call Big River. The white men call it Rio Grande (that means "big river" in Spanish).

The people of the pueblos have seen their land invaded by Spanish men and women. Their people have died mysteriously from diseases that were never known before. The Spaniards are few, but their guns and horses give them power. The Indians are forced to grow crops for the Spaniards, to pay them taxes, to clean their houses, to do their heavy work. The Spaniards take much of their land and some of their women. The pueblo people are made to worship the Spanish god. The Indians believe there is truth in all religions, so they don't mind following the ways of the Catholic church. But the Spanish priests say that only their religion is true. They call the Indian religion evil. They say that religious freedom is the freedom to worship false gods. The Spanish priests are determined to destroy the old Indian ways. Indian dances are forbidden, the masks of the Indian priests are burned, and so are their sacred kachina dolls, important in religious ceremonies. The priests report proudly to their bishops in Spain that they have destroyed 1,600 Indian masks.

The Spanish missions were small farms built around a church. Sometimes soldiers lived nearby in forts called *presidios.* Sometimes the mission and the presidio were combined. Indians lived in the missions and did most of the farming and building. The priests taught them to read and write and to become Christians.

A priest, Fray Marcos de Niza, and a black man named Estebán were the first Old World explorers in New Mexico. They were looking for gold and the seven cities of Cíbola. They didn't find either.

The Pueblo people pretend to do as the Spaniards wish, but, secretly, they keep to the old ways. They train their medicine men in kivas, dark rooms with hidden entries that are dug deep in the ground. Popé is one of those who is trained to be a leader.

The Spaniards have spies among the Indians. They learn of the medicine men and they are enraged. Spanish soldiers round up 47 Indian religious leaders. Four are accused of witchcraft and are hanged in the big plaza in the town called Santa Fe. The others are whipped and thrown in a dungeon.

Now the Indians are furious; even those who have converted to Christianity are enraged. Great numbers of Native Americans march to Santa Fe. (The Indians know the town as Bead Water.) The Spaniards have killed beloved Indian leaders; they have humiliated others. The most beloved of them all is the man named Popé.

His name means "ripe squash" in the Tewa language. We know little

The New Mexican colony got its start in 1598 when Juan de Oñate arrived from Mexico. Oñate, a Spaniard, married a woman whose grandfather was Hernando Cortés and whose great-grandfather was the Aztec emperor Moctezuma. In 1610 the settlement is moved to Santa Fe. What is going on in Plymouth in 1610? How about at Jamestown? St. Augustine?

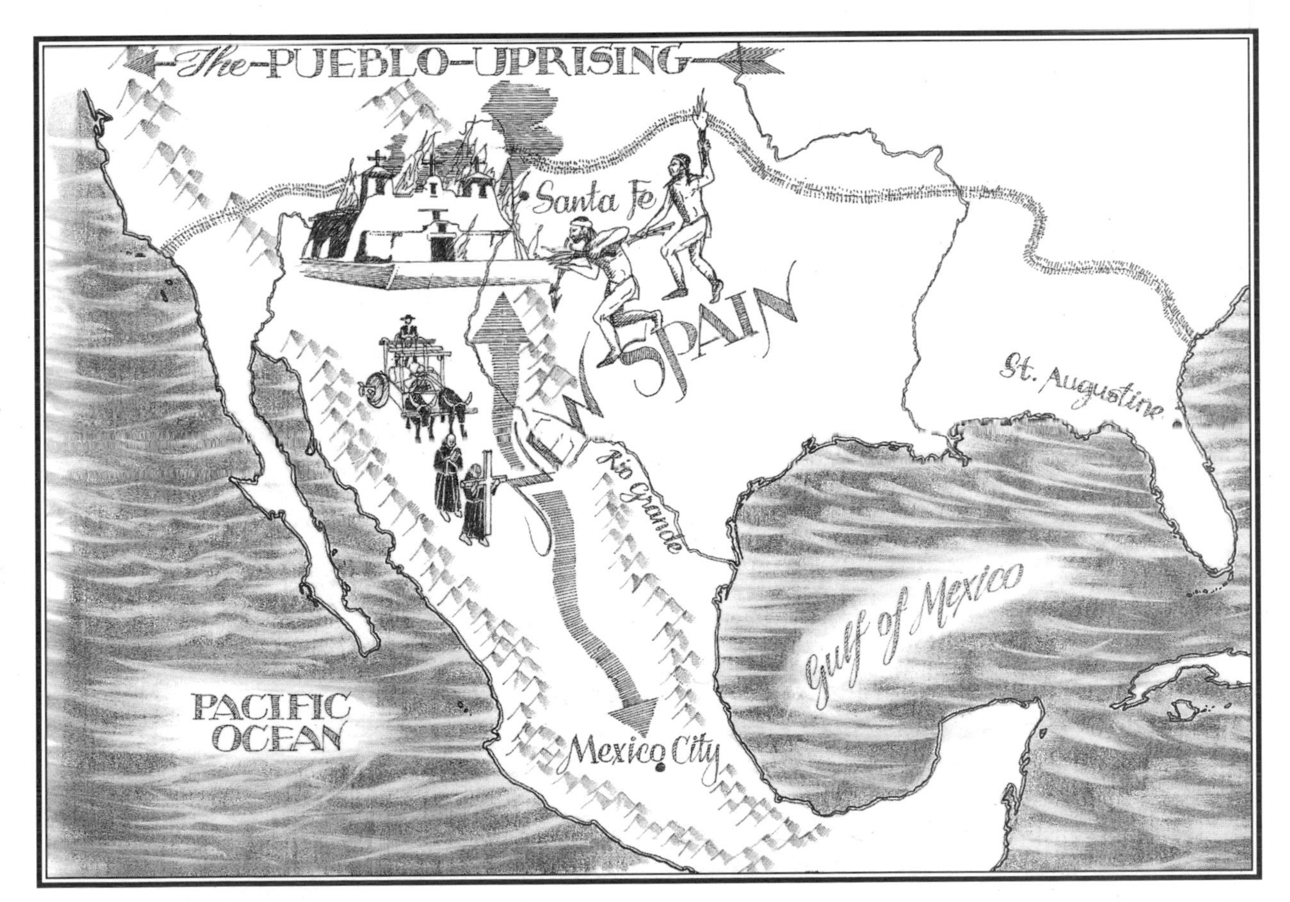

Though thousands of Pueblo Indians converted to Christianity, most still held on to their own beliefs and ceremonies, as in the dance shown here.

of him except that he was wise and good, and that he inspired others.

The Indians who march to Santa Fe say they will leave the valley forever if the men are not released. If the Indians leave, the Spaniards will have no one to work for them and they will starve. The Spaniards let Popé and all the medicine men go.

Popé has learned that the Indians have power when they unite. He must unite his people. But that will not be easy. The pueblo people of the valley speak seven different languages.

Still, he does it. Popé talks to leaders in all the pueblos. He even meets with his people's ancient enemy, the Apache. The Apaches are not pueblo people; they are nomads who have come from the Great Plains.

Nomads are people who move a lot. They settle for a while in one place, grazing their flocks. When the land is used up, or the weather changes, they move on to another spot.

Franciscans preach the gospel to two Indian leaders and their wives. One priest was asked how many "true believers" there were among his Pueblo parishioners; he said, "I don't think I have any."

The Apaches agree to help. Plans are made with great secrecy, for the Spaniards have spies everywhere.

In August 1680 Popé is ready. He knows the Spaniards have heard of the plans; he knows they are suspicious; he decides to mislead them. So he sends runners with a message—a message in a knotted rope. The Spaniards capture the runners and the Spanish governor forces an Indian to read the message of the knots. There are four knots and it means the revolution will come in four days.

That's what the message says, and that is what Popé wants the Spaniards to believe. But the next day, before they have time to prepare, all up and down the valley, at every Spanish settlement, at exactly the same time, Indians attack—burning, killing, and destroying. By nightfall only two Spanish communities are left: Santa Fe and Isleta.

Both of those towns are put under siege. That means that if anyone comes out, the Indians will overwhelm him. Because the plans have been made so well, there is no one to come to the Spaniards' rescue. The Indians have destroyed the other Spanish towns. Popé knows no help will come from Mexico. It usually takes a year and a half for wagons to make that trip. Besides, the winter's heavy rains have flooded the Rio Grande. A wagon train with supplies is stuck in the mud on the wrong side of the river.

Popé gives the Spanish governor a choice: he can stay in Santa Fe and starve, or he can leave the land and take his people with him.

The Spaniards leave—all of them. They march away, back to Mexico. The Indians watch peacefully and let them go. Then they burn the Spanish churches, tear down the Spanish towns, and go back to their old ways. They have won their revolution against tyranny and unfair taxes and religious persecution.

But the leaders in Spain and Mexico don't give up easily. They will return to New Mexico. It will take 12 years, and Popé will be dead, but they will be back. They will deal with the natives again. The Spaniards and the Pueblo Indians will do the same things that Englishmen and East Coast Indians are doing: they will fight, steal, trade, make peace, and misunderstand each other. These are civilizations in conflict. If there is a way for them live in harmony, no one seems wise enough to have found it.

In 1692 the colonial Spanish leader Don Diego de Vargas (above) and about 100 men reentered Santa Fe without bloodshed. This peace did not last. A year later, a fierce battle was fought for the city. The Pueblos were no longer united, and Vargas won. The Spaniards called it the *reconquest*.

24 What's a Colony?

Most of the colonies had no cash. They weren't supposed to mint it themselves, and England wouldn't export coins and bills. So the only way money reached the colonies was when people actually bought and sold things with it. Some colonies got so desperate for metal and paper money that they did make it themselves—or they used goods, such as tobacco, corn, and cows, instead.

Neither the landlords (Europeans) nor the renters (colonists) ever considered the Indians, who thought the land was theirs.

The Indian on the seal of the Massachusetts Bay Colony is saying, "Come over and help us"—which the Indians must have found hard to believe.

You may already know that England had 13 American colonies. I'll list them in just a minute.

First, do you remember what a colony is?

A colony is a place that belongs to another country. Think about a landlord and a renter. England, France, Spain, Holland, Sweden, and Portugal were all landlords in parts of America. They thought they owned the place. Then their colonists, the renters, said, "We're tired of belonging to someone else. We intend to own our own land." When that happened there were revolutions, and the landlords were forced to pack up and go home.

The east coast of North America was an English colony from 1607, when Jamestown began, until the Declaration of Independence was signed in 1776. Those were colonial times.

It was a long time. 1776 minus 1607 is...you can do the arithmetic yourself. When it comes to families, it was about seven generations. That means from son to father to grandfather to great-grandfather to great-great-grandfather to great-

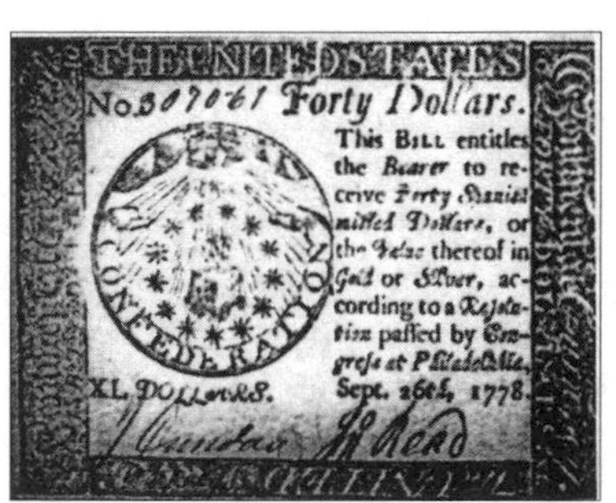

Massachusetts minted Pine Tree shillings (above). The $40 bill was backed by Spanish silver.

An artist who lived in the 19th century made this somewhat prettified portrait of an early colonial kitchen. A real 17th-century kitchen was probably barer and dirtier than this, with less furniture and fewer utensils and pots.

great-great-grandfather to—whew—great-four-times-grandfather. A lot of things can happen in seven generations.

So when you read about colonial times, don't be surprised if you read different descriptions.

When the first colonists arrived, there were no friends to greet them. No houses were ready for them. They had to start from scratch —and I do mean scratch, as in scratching. The early colonists often had to live in huts of branches and dirt, or Indian wigwams, or even caves, and none of those places was bugproof. And, of course, they had to scratch a living out of the ground.

Later colonists lived in small wooden houses with one or two rooms. Eventually, some lived in fine houses. A few lived in mansions, with beautiful furniture and paintings and dishes and silver. But no one had a bathroom like you have, or electric lights, or a furnace, or running water, or kitchen appliances. And very few people lived in mansions anyway.

You'll be reading about all 13 of the English colonies, because they

In 1633, the Dutch brought a schoolmaster to the city of New Amsterdam to teach their children. He was Adam Roelantsen (ROY-lant-sun), and he founded the first school in the North American colonies. Today that school is the Collegiate School in New York City. That same year, 1633, Boston Latin School was opened. It was the first public school in the colonies. It, too, still exists.

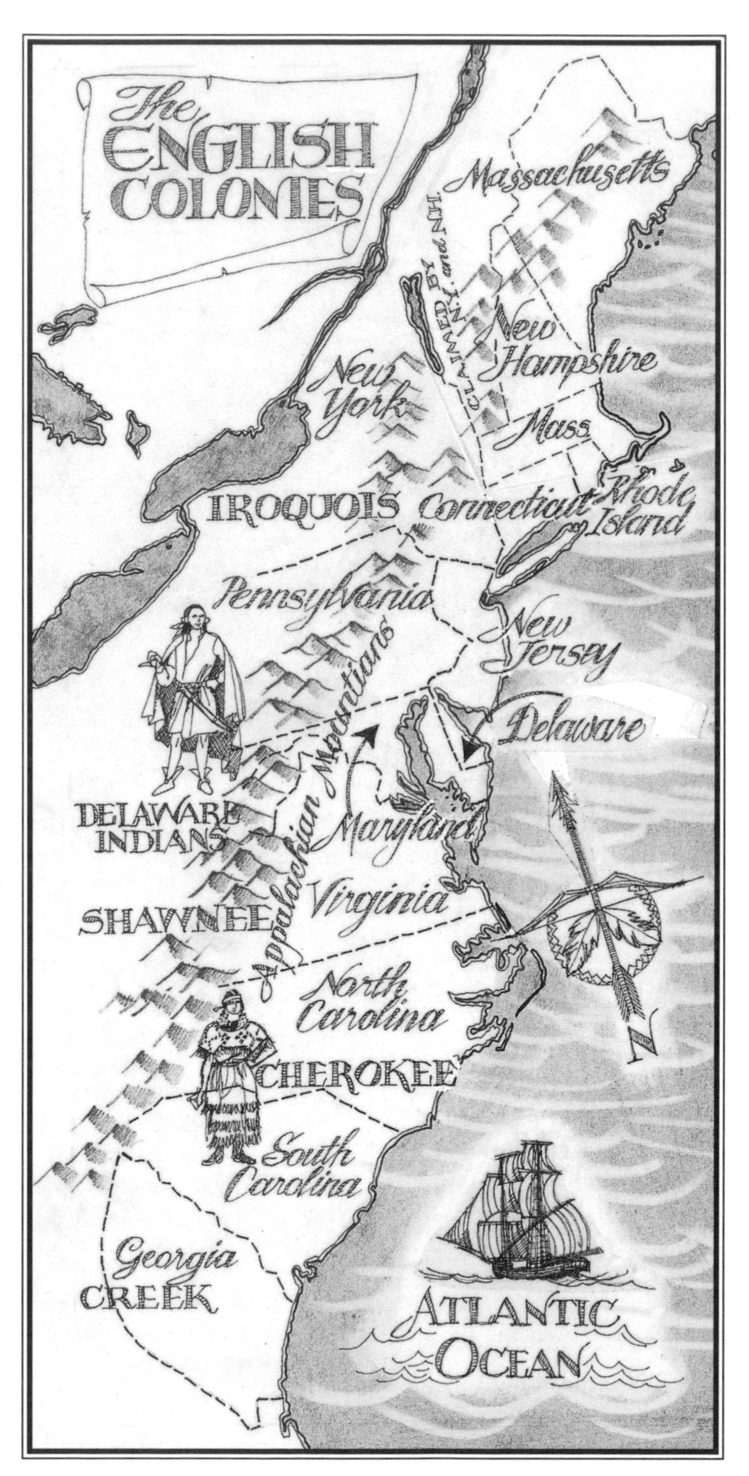

turned into the United States. To get started, we need some organization. We're going to divide the colonies into three groups: north, south, and middle.

Look at the map and you'll see the New England colonies, the Middle Atlantic colonies, and the Southern colonies.

You know a whole lot about New England, so here's a test. Name the New England colonies. (Without checking on the page opposite.)

Did you get them all?

Did you goof and say Maine? Remember, Maine was not a separate colony; it was part of Massachusetts. We haven't said anything about Vermont. Today Vermont is a New England state, and a beautiful one, but it was not a separate colony.

(Remember, don't confuse states with colonies. The 13 colonies will turn into states when the Constitution is written and our nation—the United States—is formed.)

You already know about one Southern colony, Virginia. You will learn about Maryland, North Carolina, South Carolina, and Georgia. (Some people call Maryland a middle colony, but I don't.)

Coming next in this book are the Middle Colonies: New York, New Jersey, Delaware, and Pennsylvania.

First up is New York, where the Dutch are in control. It's hard for us to realize now that tiny Holland was once a great power. The Dutch had colonies all over the world. A business firm, the Dutch West India Company, owned most of the colonies, just as the Virginia and Plymouth companies owned English colonies.

Jonas Michaelius (mick-AY-lee-uss)

Sheep were very important before the colonies grew much cotton. Sheep grow heavy coats in winter, so their wool is sheared in spring.

came to New York in 1628. It was called New Netherland then, and the Dutch West India Company was in charge. That company had promised Michaelius a home in the New World. Here is part of a letter Michaelius wrote soon after he arrived:

> *The promise which the Lords Masters of the Company had made me to make myself a home...is wholly of no avail. For their honors well know that there are no horses, cows, or laborers to be obtained here for money....The country yields many good things for the support of life, but they are all to be gathered in an uncultivated and wild state.*

The New England colonies are: Massachusetts, New Hampshire, Rhode Island, and Connecticut.

The Dutch West India Company made promises that couldn't be kept. The Virginia Company did the same thing at Jamestown. (Those early Virginia settlers really expected to find gold on the ground.)

The Puritans always told the truth about their colony. So did a man named William Penn who owned a colony. (Yes, some individuals did own colonies.) We'll get to William Penn and his colony when we finish with New Netherland.

That small Dutch colony was just a trading post. The Dutch thought that India—with its silks and spices—was much more important than America. But, just in case America did turn out to be valuable, they decided to do some fur trading on this continent.

Henry Hudson: New York's Explorer

Henry Hudson knew he had found something special on a September day in 1609 when he sailed his small Dutch ship, the *Half Moon*, into the river that would bear his name. He was looking for the Northwest Passage, and the river seemed likely. It was deep and full of salmon, mullet, and other fish.

The *Half Moon* sailed jauntily, a carved red lion with a golden mane jutting out from its forward tip. The lion was splendid, and so was the whole ship. The bow (the front part of the boat) was bright green, with carved sailors' heads in shades of red and yellow. The decks—the forecastle and the poop—were painted pale blue with white clouds. The stern (the rear of the ship) was royal blue, with stars and a picture of the Man in the Moon. That wasn't all: there were glowing lanterns and flags—the Dutch flag, the flags of all seven Dutch provinces, the flag of the Dutch East India Company, and more, too.

Now, what would you have thought if you were a Native American standing on the shore of Manhattan Island, and this great colorful seabird appeared with men standing on its back? Remember, this was 1609, and the *Half Moon* was probably the first European ship you'd ever seen. At first the Indians thought it had come from God, and that the men aboard were his messengers. It wasn't long, however, before they realized they were just men.

25 Silvernails and Big Tub

The Dutch had good relations with the Indians. One of their treaties said they would "keep the Great Chain of Friendship polished bright."

New Amsterdam's coat of arms was flanked by a pair of beavers—and fur trading paid the bills.

Old Silvernails is what they called him, because the stick of wood that stood in place of his right leg was decorated with silver nails. He had lost the leg in a battle in the West Indies. His real name was Peter Stuyvesant (STY-viss-unt), and he was a Dutch governor and a hard-swearing, tough man. Maybe the Dutch thought Silvernails Stuyvesant was the right kind of person to run a colony in America. Maybe they didn't have any other volunteers. When it came to colonial leaders, the Dutch came up with some strange men.

They had a great piece of property, but they didn't seem to realize that. They just kind of fooled around on the North American continent. They were more serious in other parts of the globe.

Back in 1609, Henry Hudson, an Englishman sailing for the Netherlands, had gone up the river that is now called the Hudson. Because of that voyage, Holland claimed a large hunk of American land. It was land wedged between the stern Puritans in the North and the Anglican tobacco planters in the South; the Dutch called it New Netherland. (Today we know it as New York and New Jersey.)

In case you're confused: Holland, the Netherlands, and the land of the Dutch are all the same place. Why don't you find it on a map? And, while you're looking, can you find England? Can you see why both nations became sea powers? Now, cross the Atlantic.

Peter Stuyvesant is still remembered in New York, where a high school is named after him.

Slave Market

In 1709, a market was set up on the corner of Wall and Water streets. It was a slave market: a place where men, women, and children were bought and sold. New York paved its streets with money raised by a tax on each slave brought into its port. By the middle of the 18th century, one in every ten New Yorkers was black. Most of them were slaves, but some were free; a few owned indentured white servants—"bound" men, women, girls, and boys.

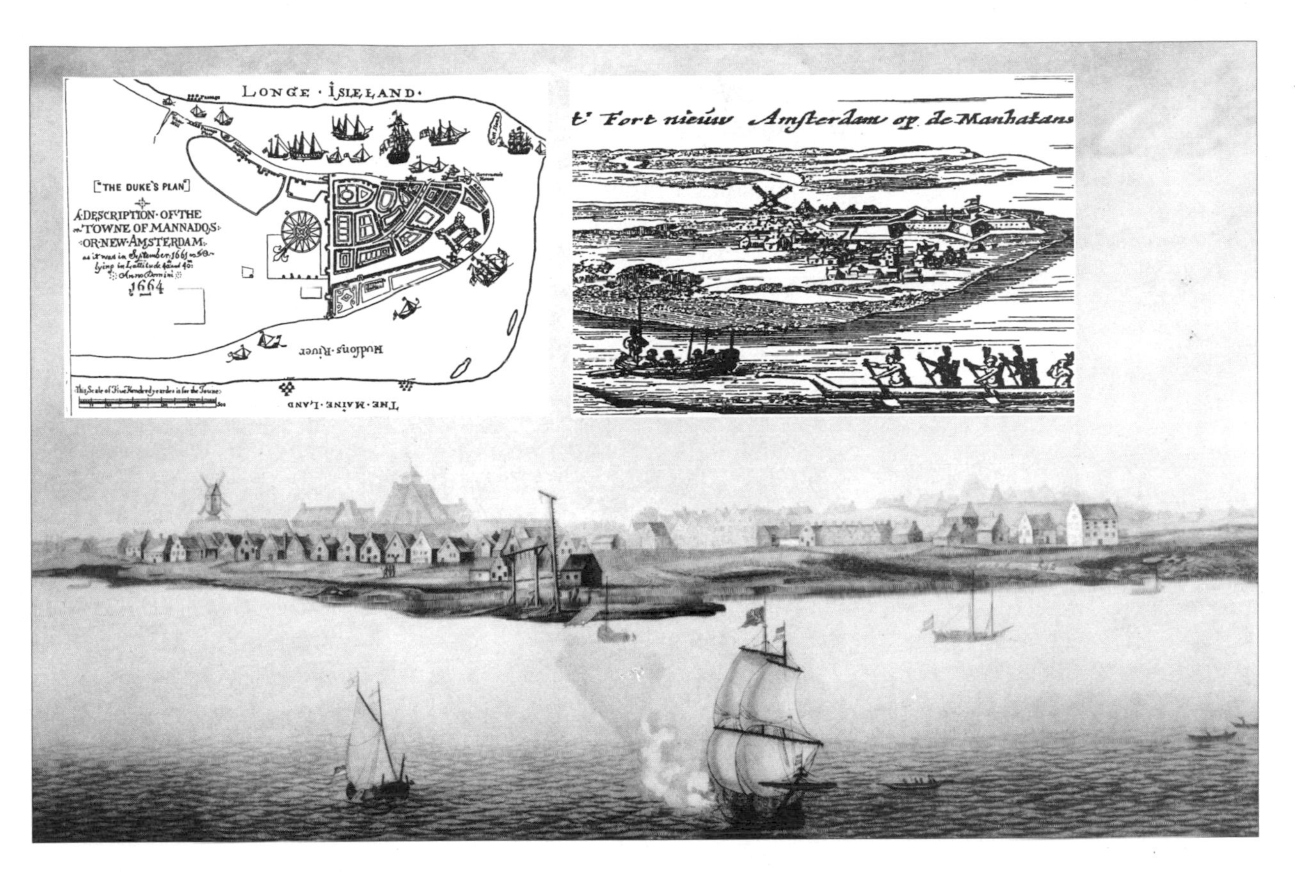

Three early views of New Amsterdam, each from a different vantage point. In the big picture the artist is looking across the East River from Brooklyn Heights, in the village the Dutch called Breukelen. The plan at top left looks down and shows clearly the line of the city wall, and the picture at top right was done on the west side of the Hudson River, in New Jersey. In the 17th century, most of Manhattan was farmland.

Put a finger on New York Harbor. Go up the Hudson River and look for a passage to China. You'll be stopped by rapids in the river when you get to Albany. That's what happened to Henry Hudson.

In 1626 the Dutch West India Company made what may be the most famous real-estate deal in history. It bought Manhattan Island (now the center of New York City) and Long Island from the Indians who lived there for some beads and goods said to be worth $24. Since the Indians didn't think people could own land, they may have thought they were outsmarting the white men.

By this time, the Dutch had decided that American furs might be almost as good as American gold, so they set up some trading posts. People in Europe were eager to buy American furs. Beaver, bear, fox, and other fur pelts could be made into sumptuous hats, coats, and blankets. The Dutch merchants hoped to get rich in the New World.

On Manhattan Island the people from Holland built a town called New Amsterdam. At one end they put a wall because they feared wolves and because cities in Europe had walls. Today that wall is the site of a famous street. Can you guess what it is called? Outside the wall were farms, which the Dutch called *bouweries*. Today a street in

You guessed it—Wall Street!

After a while, Jews in New Amsterdam were allowed to trade and own houses. But they were not allowed to build a synagogue until 1730, so services were held in people's homes.

New York is called the Bowery. (It doesn't look much like a farm now.)

Because of its great harbor, New Amsterdam was soon a sailor's town, bustling with people who arrived on ships from faraway places. It was said that you could hear 18 different languages being spoken in the city of New Amsterdam. Right away, in 1626, the Dutch brought slaves to New Amsterdam. You could buy a slave for about the same amount it would cost to pay a worker one year's salary. So some people thought it made good sense to own slaves. It was an economic, or money, decision.

One day a ship sailed into New Amsterdam's harbor with a group of Jews aboard. Peter Stuyvesant didn't want them to land. He was a bigot and didn't believe in religious freedom. Stuyvesant was a member of the Dutch Reformed Church and saw no reason to tolerate others. "Giving them liberty," he wrote ("them" meant the Jews), "we cannot refuse the Lutherans and Papists [Catholics]." But the Dutch West India Company said the Jews could stay, so there was nothing Stuyvesant could do.

Settlers in New Sweden trading with Delaware Indians. In 1643 the Swedes sold arms to the Delaware, who were fighting a Maryland tribe. Here they are making a deal to use the Indians' land for trapping furs.

Grouchy Old Silvernails was in charge of all of New Netherland. He was pretty good at running things, but he would stomp his wooden leg and swear at anyone who disagreed with him. When the Dutchmen who had been elected councilors objected to something he said, he called them "ignorant subjects." Another time he said he would ship them back to Holland—in pieces—if they gave him trouble.

But if you were going to run a swearing contest, Stuyvesant might lose. Johan Printz (YO-han PRINCE) had a mouth that was even more foul. He was governor of New Sweden, on the Delaware Bay, not far from New Netherland.

Printz was a whale of a man. The Indians called him Big Tub, and he may have been the biggest man on the continent. He was seven feet tall and weighed 400 pounds. Big Tub was an autocrat—an absolute ruler—and he liked to hang people who opposed him. But he did hold his colony together for 10 years, with very little help from the Swedes at home. And he introduced a new style of architecture, the log cabin, that became popular in frontier settlements.

Finally, Johan Printz got tired of trying to run things himself and went back to Sweden. The new Swedish governor decided to get tough and capture a Dutch fort, but he didn't realize what Old Silvernails was like. Stuyvesant sent seven ships, and that wiped out the Swedish colony.

That made Stuyvesant popular in New Amsterdam—but not for long. Nine years later, in 1664, the English decided to do to the Dutch what the Dutch had done to the Swedes. Old Silvernails stomped on his wooden leg, but nobody came to his rescue. The British took New Amsterdam without firing a shot. They renamed it New York.

Manhattan is an island 12½ miles from end to end. Broadway, its longest street, runs the entire length of the island.

The Hon. Peter Stuyvesant
a Pennsylvania Bank Barn
Printing Press
Wooden Rocking Horse
Ice Skates
Small Windmill
Indian corn
Alcove Bed
Crabs
and Tavern signs.
1794
J. VANDER HAYDEN
ENTERTAINMENT
The MIDDLE COLONIES
St. Lawrence R.
Lake Ontario
CLAIMED BY N.Y. & N.H.
Hudson R.
Albany
NEW YORK
PENNSYLVANIA
Susquehanna R.
Delaware R.
New York
Long Island
NEW JERSEY
Atlantic Ocean
Lancaster
Philadelphia
DELAWARE
N
W
E
S

26 West to Jersey

James, Duke of York, later became James II, king of England, but he was not popular.

The Netherlands pulled down its flag and le America. It was 1664, and the Dutch had bee here for 50 years. During those years they ha gone up the Hudson River, built a town at For Orange (Albany), and established big planta tions along the river. The Dutch farm owner were called *patroons*. People came fro Holland to work the patroons' farms. The brought their tulips, their hardworking habit their neatness, and their storytelling ways wit them. (Their children's favorite story was about a fellow named Sant Claus, who visited just before Christmas, on December 6.)

Actually, most of the Dutch people didn't leave, and the Britis even let Peter Stuyvesant stay. They let the the Dutch have religiou freedom. The Netherlanders just weren't in charge anymore. It wa the Duke of York who owned the place now.

The Duke of York was named James, and his brother was the Kin of England. York (that is what some people called James) was one the owners of the Royal African Company, which controlled th British slave trade. It seems to have occurred to the duke that if he en couraged the use of slaves in the New World, he would make lots money. And that is just what happened.

York must have been conceited and vain, because you can se what he named his land. Well, that isn't quite true. It wasn't all calle New York. Some of it became New Jersey. The Duke of York gave a bi chunk of land to two of his friends. One of them, Sir George Cartere came from the island of Jersey, which explains why it was "New Jersey. The duke's other friend was Lord John Berkeley. Berkeley' brother was governor of Virginia.

Those two owners were called proprietors. They expected the pe

The English came up with the name Holland, meaning "hollow land." They were thinking of the center of the country, which is unusually flat. Much of Holland's land was once under water. Beginning in the 12th century, the Dutch built dikes, which are sea walls. Then they drained the land. The dikes kept the ocean from returning. Do you know the story of the boy who put his finger in a hole in a dike and saved his people?

Fort Orange, which became Albany, was established as a fur trading post before New Amsterdam.

ple who lived on their land to pay them a tax called a *quit rent*. That didn't make them very popular.

But they had good qualities all the same. They wrote a plan of government, a charter, that was the best any English colony had. It set up an assembly that represented the settlers. (Assemblies, parliaments, and congresses are all similar organizations.) The charter provided for freedom of religion. You could be a Quaker or a Puritan or an Anglican in New Jersey, and, as long as you were a man, you had the right to vote. You didn't have that freedom in Massachusetts or Virginia or most of the other colonies.

Soon people were pouring into New Jersey from all over: Finns, Swedes, Germans, English and others. In New Jersey everybody lived together in harmony.

At first New Jersey was divided into East Jersey and West Jersey. Then the king bought out the proprietors, united New Jersey, and made it a royal colony. That meant the king was now the owner of New Jersey. He sent a royal governor to take charge and collect his rents. The king allowed the colonists to keep making their own laws through the elected assembly.

For a while, New Jersey was part of the western frontier of the country. The frontier was land that was on the edge of what Europeans considered civilization. If you look at a map, you can see where the frontier of European civilization was in 1700. Is there a frontier today? Where might it be? Clue: look up.

These are rows of old Dutch houses in Albany. You can tell Dutch houses by the roofs with their "steps," called gables.

27 Cromwell and Charles

On January 30, 1649, just before he was beheaded, Charles I asked for a warmer shirt than usual, lest the cold make him tremble as though from fear. He said, "I fear not death. Death is not terrible to me. I bless my God I am prepared." When the executioner held up the head of the king, the great crowd gave out a groan. "It was such a groan," said a witness, "as I never heard before, and desire I may never hear again."

Charles didn't have to lose his head, but he wouldn't give an inch to his opponents. That was his downfall.

Even though this is a book about US, you need to keep up with events in England, because what was going on there was very important to the colonists in America. And something unexpected was happening in England. A civil war was being fought. The war was between King Charles I and the Puritans. The king lost the war. Then the Puritans executed him. (Kings are never killed, they are executed.)

Are you surprised about the civil war? Are you surprised that the English would chop off their king's head? Back then a lot of people were astonished.

Some people called the new government the Protectorate. Some people called it the pits.

Virginians were on the side of the king; people on that side were called Cavaliers. (That name was because the king's soldiers fought mostly on horseback, and soldiers on horseback are called cavalry.)

The New Englanders, naturally, were for the Puritans, who were sometimes called Roundheads. (They wore their hair short and rounded at a time when many men wore long, powdered wigs. When they needed a haircut, Puritan men would put a bowl on their heads and cut around the bowl, like the picture in chapter 21.)

Oliver Cromwell (right) was a very good general who often beat the Royalists even when he was outnumbered. He wasn't a very good ruler, unfortunately.

Put 1649 in your head. Because—talking about heads—1649 was the year Charles I lost his. It was an important act in history because it reminded

people that kings could be overthrown. And, even though it happened in England, it helped bring freedom to Americans.

That revolution in 1649—and it really was a revolution—was called the English Civil War. I'll say it again, because it is so important: In 1649 the Puritans won the English Civil War, and Charles I lost his head.

Some people think this revolution was a failure—because, as you soon learn, kings came back. But it wasn't a failure. The Civil War made an important point: people can change their government if they want to badly enough.

Now, back to 1649. Since poor King Charles I was without a head, the Puritan leader, Oliver Cromwell, took charge of the government.

At King Charles's trial, the judge put on a scarlet gown to read the death sentence: "Charles Stuart is a tyrant, traitor, murderer, and public enemy to the good people of this nation, and shall be put to death by the severing of his head from his body."

The first Cavaliers were swaggering young men who acted as Charles's personal guard.

I would like to tell you that Cromwell did a splendid job of ruling, but he didn't.

When the Puritans got power, they destroyed many Anglican churches, broke church windows, smashed statues, and burned great works of art. They even killed some Catholics and Anglicans, just because of their religion.

Oliver Cromwell tried to do what he thought was right. He did many good things, and he was a strong leader. But he didn't understand that his opponents had rights, too. Some people called him a tyrant.

One of the first things the Puritans did was to close all the theaters in London. That was going too far. Some of the plays in London seemed wicked, but the Puritans wouldn't even let Shakespeare's plays be seen.

The Puritans didn't seem to have much of a sense of humor. In one of Shakespeare's plays, called *Twelfth Night*, Sir Toby Belch, a clownlike character, makes fun of Puritans. Sir Toby says, "Dost thou think, because thou art virtuous, there shall be no more cakes and ale?" (What did he mean by that?)

When Cromwell died, his son Richard—who was known as Tumbledown Dick—tried to take over. Dick was a quiet man who liked farming better than ruling. He couldn't hold on to power. Besides, the English people were tired of the stern Puritans, and they wanted to have fun. They wanted cakes and ale. So they put the old king's son on the throne. He was Charles II, and because of his good nature he was known as the Merry Monarch. The time in which he ruled is called the Restoration. (You can see why it was the Restoration—kings were *restored* to power.)

But Charles wasn't so merry when it came to the Puritans. He killed some of them, and he dug up Cromwell's body and cut off his head. Now Puritans were being mistreated. Things got so bad during the Restoration that many Puritans left England and came to America.

Some of the king's friends came, too. When Charles II became king, he decided to reward friends who had stayed loyal to his family while the Cromwells were running things. He gave them gifts of America. As you know, he gave his brother, the Duke of York, the gift of New Netherland. (Then he sent an army to take it from the Dutch.)

He gave the Carolinas to eight of his favorite lords. He gave Pennsylvania to a young man whose religion made it dangerous for him to live in England. The next chapter will tell you why.

The Plague that devastated London in Charles II's reign ended partly because of another disaster—the Great Fire of 1666. It started in a bakery, and when it was over most of London's old wooden houses had gone. But so had the Plague germs.

28 William the Wise

You've probably noticed that the spelling of some of the original documents quoted in *A History of US* isn't the same as the spelling you've learned. Until Noah Webster came along with a speller (in 1783) and an American dictionary (in 1828), there was no standard spelling. People just wrote words the way that seemed right to them. Sometimes the same word could be spelled two different ways in a single sentence.

"I have led the greatest colony into America that ever any man did upon a private credit."

William Penn was born with a silver spoon in his mouth and servants at his feet. His father was an important admiral: rich, Anglican, and a friend of King Charles II.

What did William Penn do when he grew up? He became a member of a radical, hated, outcast sect, the Society of Friends, also known as the Quakers.

What did being a Quaker do for William Penn? It got him kicked out of college when he refused to attend Anglican prayers. It got him a beating from his father, who wanted him to belong to the Church of England. It led him to jail for his beliefs—more than once. It gave him a faith that he carried through his life. And it also gave him a reason for founding an American colony.

King Charles II liked William Penn in spite of his religion. Everyone, it seems, was charmed by his sweet ways. But when Penn came before the king and refused to take off his hat—Quakers defer only to God—some people gasped and wondered if Penn's head, along with his hat, might be removed. But Charles, the "merry monarch," must have been in a good mood. As the story goes, he laughed and doffed his own hat, saying, "Only one head can be covered in the presence of a king."

Now King Charles had borrowed money from Admiral Penn, and a goodly sum it must have been, because, after the admiral died, when William asked that the debt be paid with land in America, he was given a tract of land larger than all of England. King Charles named it Pennsylvania, which means Penn's woods.

Pennsylvania was situated midway between the pious Puritans in

An Hiſtorical and Geographical Account
OF THE
PROVINCE and COUNTRY
OF
PENSILVANIA;
AND OF
Weſt-New-Jerſey
IN
AMERICA.

The Riehneſs of the Soil, the *Sweetneſs* of the Situation, the Wholeſomneſs of the Air, the Navigable Rivers, and others, the prodigious Encreaſe of Corn, the flouriſhing Condition of the City of *Philadelphia*, with the ſtately Buildings, and other Improvements there. The ſtrange Creatures, as *Birds, Beasts, Fishes,* and *Fowls*, with the ſeveral ſorts of *Minerals, Purging Waters,* and *Stones*, lately diſcovered. The *Natives, Aborogmes*, their *Language, Religion, Laws*, and *Cuſtoms*; The firſt Planters, the *Dutch, Sweeds*, and *Engliſh*, with the number of its Inhabitants; As alſo a Touch upon *George Keith's New Religion*, in his ſecond Change ſince he left the *QUAKERS*

With a Map of both Countries.

By GABRIEL THOMAS, who reſided there about Fifteen Years.

London, Printed for, and Sold by *A. Baldwin*, at the *Oxon Arms* in *Warwick-Lane*, 1698.

Gabriel Thomas, one of the earliest settlers in Pennsylvania, wrote a pamphlet in praise of its charms—you can read his words on page 104.

New England and the convivial Anglicans in the South. Quakers weren't wanted in either region.

Thanks to William Penn, Quakers now had their own colony. But he made it different from most of the other colonies. Penn really believed in brotherly love. He said that Pennsylvania was not just for Quakers but for everyone.

The king had picked a good man to lead a colony—perhaps the best of all who tried it. Penn was an educated man, a philosopher, a town planner, and a lawyer. He wanted Pennsylvania to be a colony where Quaker ideas about peace and goodness would prevail.

Charles II died a Catholic at heart, but he admired William Penn enough to permit him to practice his religion in America.

In England, Quakers seemed a threat to everyone who felt comfortable with the old, established ways of thinking. The country had beheaded a king, and that didn't work out. New ideas seemed dangerous, as they often do. Quakers had notions that would change Old England. Wealthy citizens didn't want things to change, so it was poor people, mostly, who were Quakers.

In Penn's day, some people—ministers, kings, lords, and dukes—were considered superior to the average person. They expected others to bow to them, but Quakers wouldn't. They wouldn't bow to anyone. They even refused to pay taxes to support the Church of England. Can you see a problem? The Anglicans did.

England had lords and ladies in the rich upper class, merchants and farmers in the middle class, and peasants and poor people in the lower class. It was almost impossible to rise from the lower class to the upper. The upper-class lords and earls often acted as if they were better than anyone else. It was that class system that made many ambitious people come to the New World. In America, with hard work, many poor people would rise to the top.

Because the Bible says, "Thou shalt not kill," Quakers believe all war is wrong. They won't fight even when drafted into the army. They are called conscientious objectors, because their conscience tells them not to fight.

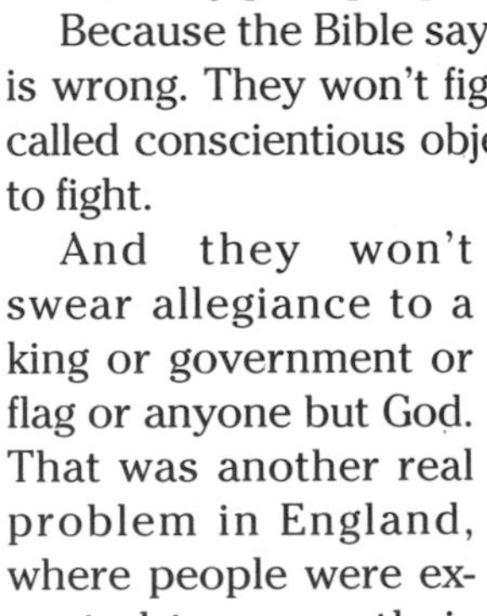

And they won't swear allegiance to a king or government or flag or anyone but God. That was another real problem in England, where people were expected to swear their loyalty to the king.

William Penn wanted to practice Quaker

Part of Penn's 1699 house in Philadelphia is still standing.

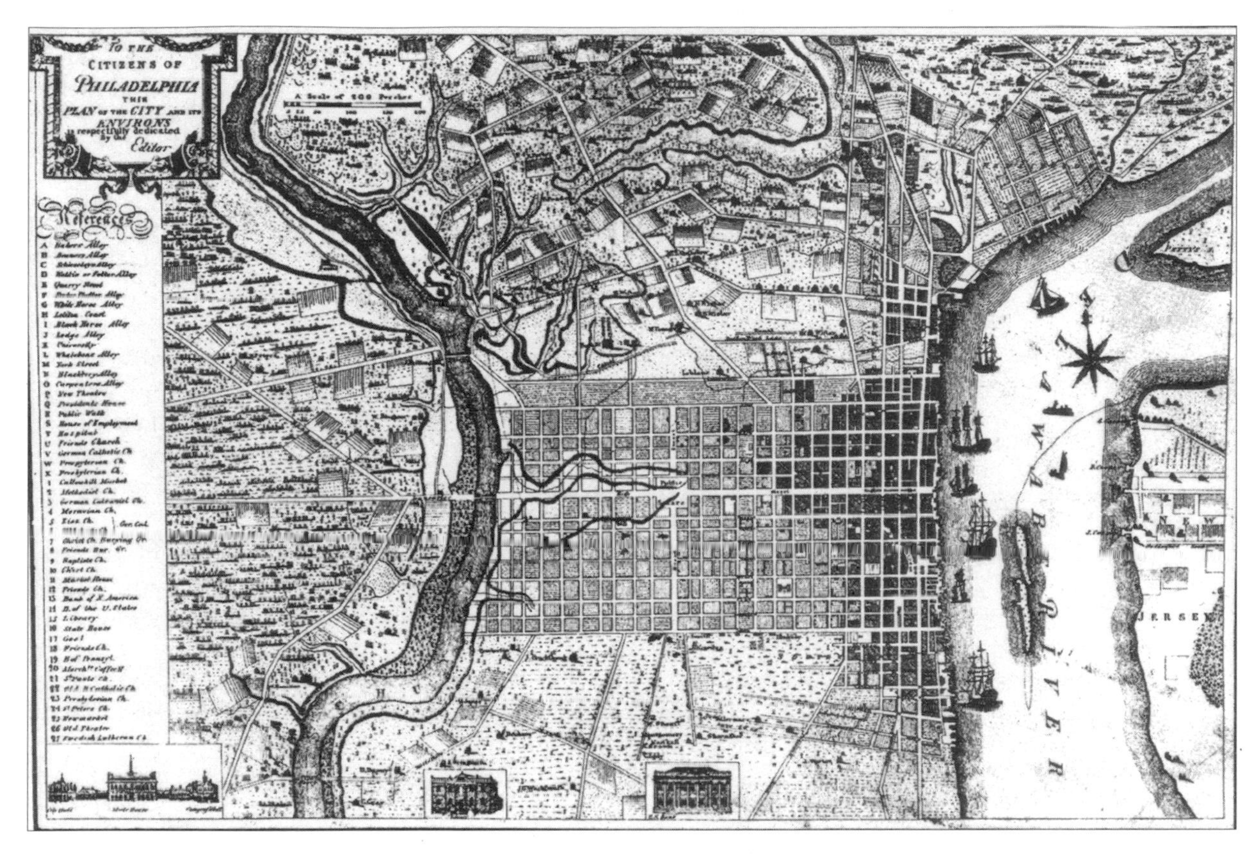

ideas in America. That meant treating all people as equals and respecting all religions. Those new ideas of "toleration" and "natural rights" were confusing. It was difficult for good people to know what was right.

William Penn planned his city, Philadelphia, without walls or fortifications because he expected its citizens to be peaceful.

Do you understand the difference between toleration and equality? Some colonies offered freedom of religion but not equality. You could practice any religion but you couldn't vote or hold office (be a mayor or sheriff) unless you belonged to the majority's church. That wasn't true in Penn's colony. While he was in charge, all religions were equal.

When Penn said all people, he meant *all* people. Quakers were among the first to object to Negro slavery and, more than anyone else, to treat Indians as equals. In 1681, William Penn wrote a letter to the Native Americans of Pennsylvania. He said:

> *may [we] always live together as neighbors and friends, else what would the great God say to us, who hath made us not to devour and destroy one another, but live soberly and kindly together in the world?*

Penn proposed a "firm league of peace." He continued:

> *I am very sensible of the unkindness and injustice that hath been too much exercised toward you by the people of these parts of the*

Not Fiction or Flam

Poor people (both Men and Women) of all kinds, can here get three times the Wages for their Labour they can in England or Wales....Reader, what I have here written, is not Fiction, Flam, Whim, or any sinister Design...but in meer Pity and pure Compassion to the Numbers of Poor Labouring Men, Women, and Children in England, half starv'd, visible in their meagre looks, that are continually wandering up and down looking for Employment without finding any....Here are no Beggars to be seen.... Jealousie among Men is here very rare...nor are old Maids to be met with; for all commonly Marry before they are Twenty Years of Age and seldom any young Married Woman but hath a Child in her Belly, or one upon her Lap.

What I have deliver'd concerning this Province, is indisputably true, I was an Eye-Witness to it all, for I went in the first Ship that was bound from England for that Countrey, since it received the name of Pensilvania, which was in the Year 1681.

—From Gabriel Thomas's Historical and Geographical Account of Pensilvania *(1698).*

world...but I am not such a man...I desire to win and gain your love and friendship by a kind, just, and peaceable life.

Penn was generous as well as fair. He offered land on easy terms to those who came to his colony.

On his first visit to America, he sailed up the Delaware River and picked the site of Pennsylvania's first capital, Philadelphia. Then he helped plan the city by using a pattern of crossing streets, called a grid, that would be copied throughout the new land. He gave numbers to all the streets that went in one direction; the streets that went the other way he gave tree names, like Pine and Chestnut and Walnut. Philadelphia is still thought of as a fine example of town planning.

Penn wrote a Charter of Liberties for Pennsylvania. Penn said the charter set up a government "free to the people under it, where the laws rule, and the people are a party to those laws."

The southeastern part of Pennsylvania was called the Three Lower Counties. In 1704 those counties asked for their own assembly and William Penn gave it to them. In 1776 they became an independent state named Delaware.

William Penn didn't stay in America for long. He had business to attend to in England, and so he chose rulers for Pennsylvania. Since he owned the place, he had a right to do that.

Penn did not believe in democracy. (Hardly anyone did at the beginning of the 18th century.) He was an aristocrat. In those days, ordinary people were not thought to be capable of picking their own leaders. William Penn thought he was choosing good people to lead his colony. But, as it turned out, he was too trusting.

The men he picked to run his colony fought among themselves and cheated him. (He would have been better off if he had believed in democracy.) William Penn lost most of his fortune developing Pennsylvania.

But Penn did prove that freedom and fairness work. Philadelphia was soon the largest and most prosperous city in the colonies. People came from Germany, France, Wales, and Scotland—as well as England—looking for religious freedom and a good place to live. One boy, named Benjamin Franklin, came from Boston.

William Penn and the Indians make peace.

29 Ben Franklin

Franklin was famous for his countrified appearance. The French couldn't get over his fur hat.

Some people had problems with Benjamin Franklin. They accused him of not having any gravity. Now that doesn't mean he floated around like a weightless space voyager. Gravity has another meaning, as in "grave." No, not a place where you get buried, but you are getting closer. Someone who is grave is very serious, maybe a bit dull, and certainly not much fun. Ben Franklin did have a problem. He just couldn't stay serious or dull. He was always playing jokes or having fun.

The French had no trouble with Ben. They loved his jokes and admired his good mind. They were amazed by all the things he had done. He was a scientist, an inventor, a writer, and a great patriotic American.

His mind never seemed to stop for rest. Daylight-savings time was his idea; and he invented bifocal glasses, the lightning rod, the one-arm desk chair, and an efficient stove. He founded the first public library, the first city hospital, and the University of Pennsylvania. He was the most famous journalist of his time, and the first editor to use cartoons as illustrations. He made electricity into a science. And that is only part of what he did.

Benjamin Franklin helped with the ideas that made this country special, and he got the French to help pay for the revolution that made us free.

But, as we said, some people had problems with Franklin. The English people didn't much like him. Well, that's not quite true. It was English politicians who didn't like him, especially when the colonies began to object to the way England was treating them.

When Ben was sent to London to represent the Americans, one Englishman wrote, "I look upon him as a dangerous engine." And Lord

Vive Franklin!

The French were fascinated by Benjamin Franklin. (Vive means "long live" in French.) One Frenchman who met him at court said:

The most surprising thing was the contrast between the luxury of our capital, the elegance of our fashions, the magnificence of Versailles...the polite haughtiness of our nobility—and Benjamin Franklin. His clothing was rustic, his bearing simple but dignified, his language direct, his hair unpowdered. It was as though classic simplicity, the figure of a thinker of the time of Plato...had suddenly been brought by magic into... the 18th century.

This poem was written in Franklin's honor:

To steal from Heaven its sacred fire he taught;
The arts to thrive in savage climes he brought;
In the New World the first of Men esteem'd;
Among the Greeks a God he has been deem'd.

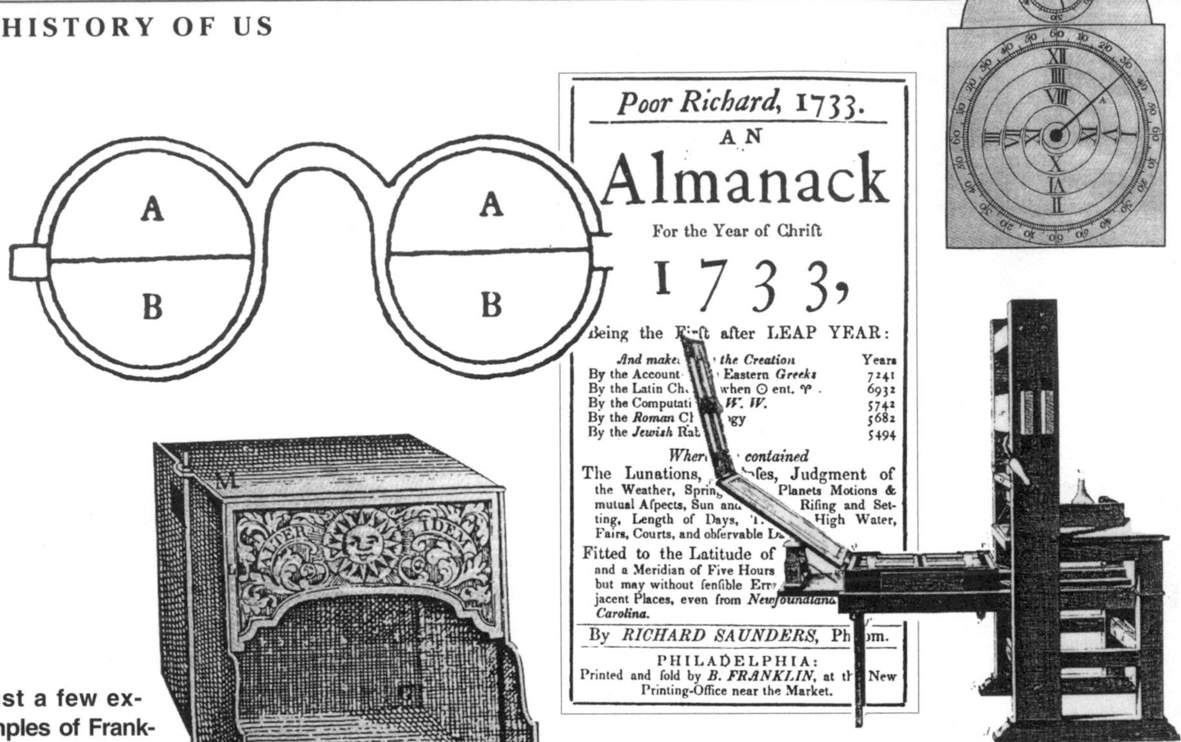

Just a few examples of Franklin's many interests. Clockwise from lower left: a Franklin stove, bifocals, a copy of *Poor Richard's Almanack*, a two-faced clock (the lower part tells hours and minutes, the upper part tells the seconds—but it was too hard to read and never caught on); and, of course, a printing press.

Sandwich (that really was his name) called him one of the "most mischievous enemies" that England ever knew. (By the way, sandwiches were named for Lord Sandwich, whose real name was John Montagu. He liked his servants to fix him a snack—meat between bread—when he played cards.)

Some Americans accused Franklin of liking the English too much; the English, of course, said he didn't like them at all. What Ben was doing was trying to be fair and also trying to prevent war. He said, "There never was a good war or a bad peace."

I think you would have liked him. And if you want to have a real hero, someone to use as a guide for ordering your life, you can't do better than Benjamin Franklin. He had what has been called a "happy balance of earnestness and humor." He made the most of what he had.

One of his biographers wrote, "He had a talent for happiness." Another said, "He hated solemn pompous people....He gave away much of his money...he set about improving himself."

But we need to tell you something of his life, so you can judge the man yourself.

Benjamin Franklin was born in Boston, the 15th child in a family with 17 children. He was the youngest son of the youngest son of the youngest son—back to his great-great-grandfather. His father was a

Benjamin Franklin was born on January 17, 1706, and died on April 17, 1790. His life spanned the 18th century.

hardworking candlemaker descended from Puritan stock. Young Franklin went to school for three years, and then his parents could afford no more. It was enough to get him started; he loved books and reading, and he educated himself.

One thing he didn't like was candlemaking. So his father signed him as an apprentice to Ben's older half-brother, James, who was a printer. In return for room and board (food) and training, an apprentice had to work for a certain number of years. He was not free to quit. It was something like being an indentured servant. Ben was 12 years old, and his father signed him for nine years.

He didn't get along with his brother, and what he really wanted to do was to go to sea. But he made the best of a bad situation. The print shop was a good place to learn. There were stacks of books—he read them all—and interesting people dropped by. Some wrote for the newspaper James printed and owned.

Ben wanted to write for the paper, but he knew James wouldn't publish his work. So he wrote letters to the editor and signed them with the made-up name of a woman; he called her Silence Dogood.

The earliest portrait of Ben Franklin in existence was painted in about 1748, when he was 42 years old and the postmaster of Philadelphia.

The letters were a big hit. Silence wrote that she was "naturally very jealous for the rights and liberties of my country: and the least appearance of an incroachment on those invaluable privileges, is apt to make my blood boil exceedingly."

When Dogood wrote a description of herself, it could have been a description of Franklin: "I never intend to wrap my talent in a napkin," she wrote. "To be brief; I am courteous and affable, good-humoured (unless I am first provoked), handsome, and sometimes witty."

Everyone wanted to know who Silence Dogood was. When Ben's brother found out, he stopped printing the letters.

Ben wasn't happy, but he didn't sit around and mope. One thing Benjamin Franklin did all his life was to try and find ways to improve himself. Maybe it was his Puritan background that made him industrious.

He decided he wanted to be a good writer. Ben had learned to

Incroachment (which we spell *encroachment*) means infringing or crossing the boundaries of something.

That's Franklin on the horse in the middle. He's being honored in a parade because he helped fight off Indian attacks on the Pennsylvania frontier in 1755.

spell at the printing house, but his father told him that the style of his writing was not good. So he found a friend and they wrote letters back and forth. Then he worked out exercises to improve his writing style. Sometimes he turned stories into poems, then back again into prose. It was a fine way to learn to work with words.

When Ben was sixteen, he read about a vegetable diet. He became a vegetarian and bought books with the money he saved by not eating meat. Soon he could talk about books with anyone. He was becoming very well educated.

But he still had problems with his brother. "Perhaps I was too saucy and provoking," said Ben. "My brother was passionate, and had often beaten me, which I took extremely amiss." Finally, at 17, Ben ran away from Boston. He sold his books and used the money to get to Philadelphia. In his *Autobiography*, Franklin described his arrival in that city:

> *I was dirty from my journey; my pockets were stuffed out with shirts and stockings; I knew no soul nor where to look for lodging. I was fatigued with traveling, rowing, and want of rest; I was very hungry; and my whole stock of cash consisted of a Dutch dollar and about a shilling in copper. The latter I gave the people of the boat for my passage, who at first refused it on account of my rowing; but I*

Saucy means fresh. A ***shilling*** was an English silver coin, worth one twentieth of a pound.

insisted on their taking it, a man being sometimes more generous when he has but a little money than when he has plenty, perhaps through fear of being thought to have but little.

Then I walked up the street, gazing about, till near the market house I met a boy with bread. I had made many a meal on bread, and inquiring where he got it, I went immediately to the baker's he directed me to, in Second Street, and asked for biscuit, intending such as we had in Boston; but they, it seems, were not made in Philadelphia. Then I asked for a three-penny loaf, and was told they had none such. So, not considering or knowing the difference of money, and the greater cheapness nor the names of his bread, I bade him give me three-penny-worth of any sort. He gave me, accordingly three great puffy rolls. I was surprised at the quantity, but took it, and, having no room in my pockets, walked off with a roll under each arm, and eating the other. Thus I went up Market Street as far as Fourth Street, passing by the door of Mr. Read, my future wife's father; when she, standing at the door, saw me, and thought I made, as I certainly did, a most awkward, ridiculous appearance.

That awkward boy soon had a job working for a printer. Then he got a chance to go to London, and went. And then he was back in Philadelphia and opened his own print shop and started his own newspaper. He was such a good and careful printer that he printed all of the colony of Pennsylvania's official papers. He was the best printer in Philadelphia.

Then he decided to publish an almanac. In the 17th century, almost every colonial home had two books: one was a Bible, the other was an almanac. The almanac was very useful. It had a calendar; it predicted the weather; it told when the moon would be full and when it would be a sliver; it told about the ocean's tides, and it was filled with odds and ends of information. Ben Franklin, with his curiosity, had a head filled with interesting information. His almanac, called *Poor Richard's Almanack*, became very popular. Poor Richard was always giving advice. He said things like:

Early to bed and early to rise, makes a man healthy, wealthy and wise.
God helps them who help themselves.
Three may keep a secret, if two of them are dead.

The almanac helped make Benjamin Franklin rich. He was able to retire at 42 and spend the rest of his life doing all the things he wanted to do. He studied electricity, invented things, became a politician, and was soon Philadelphia's most famous citizen. He would help found a new nation. By his example he showed that in America a poor boy could become the equal of anyone in the world.

What did Ben Franklin say when he discovered electricity?

Nothing. He was too shocked.

If you would not be forgotten,
As soon as you are dead and rotten,
Either write things worth reading,
Or do things worth the writing.
Benjamin Franklin

This was Benjamin Franklin's proposal for an experiment to decide whether lightning is electricity. "On the top of some high tower or steeple, place a kind of sentry-box...big enough to contain a man and an electric stand. From the middle of the stand let an iron rod rise and pass bending out of the door, and then upright 20 feet, pointed very sharp at the end. A man standing on it, when such clouds are passing low, might be electrified."

30 Maryland's Form of Toleration

Cecil Calvert, Maryland's owner, could make his own coins—here's his face on a 1659 silver shilling.

Did you notice that we have been traveling south? After New England we stopped in New York and New Jersey and then Pennsylvania and Delaware. Now, if you look around, you'll see we've reached fertile, water-lapped Maryland.

The man who got Maryland started was not at all like Ben Franklin. He was not a poor boy who had to do everything for himself; he was wealthy, very wealthy. He was also energetic and daring.

Sir George Calvert was an English lord and a real gentleman. That means he acted like one.

Sir George made things hard for himself by becoming a Catholic. The English didn't like Catholics.

But, remember, the Spaniards, French, and Portuguese didn't like Protestants. This may seem a bit tiresome, but it was serious business back in the 17th century. You could get your head cut off if you practiced the wrong religion.

Sir George didn't get his head cut off, but he was forced to resign from his important government position. Since everyone liked him (he was a real gentleman), things worked out. The Irish—who were Catholic—were happy when the king gave him land in Ireland and named him Baron of Baltimore. Then King Charles I, who really liked him, gave him a colony in America.

They named it Maryland. That was said to be in honor of the king's wife, Queen Henrietta Maria. Some say Calvert really meant to honor Mary, the mother of Jesus Christ.

Sir George dreamed of a colony "founded on religious freedom, where there would not only be a good life, but also a prosperous one, for those bold enough to take the risk."

It was George Calvert's son, Cecil (SESS-ul), who actually founded

Maryland is sometimes called a Southern colony and sometimes a Middle colony. During the Civil War it was a "Border colony." Can you guess why Maryland has had a hard time with its identity?

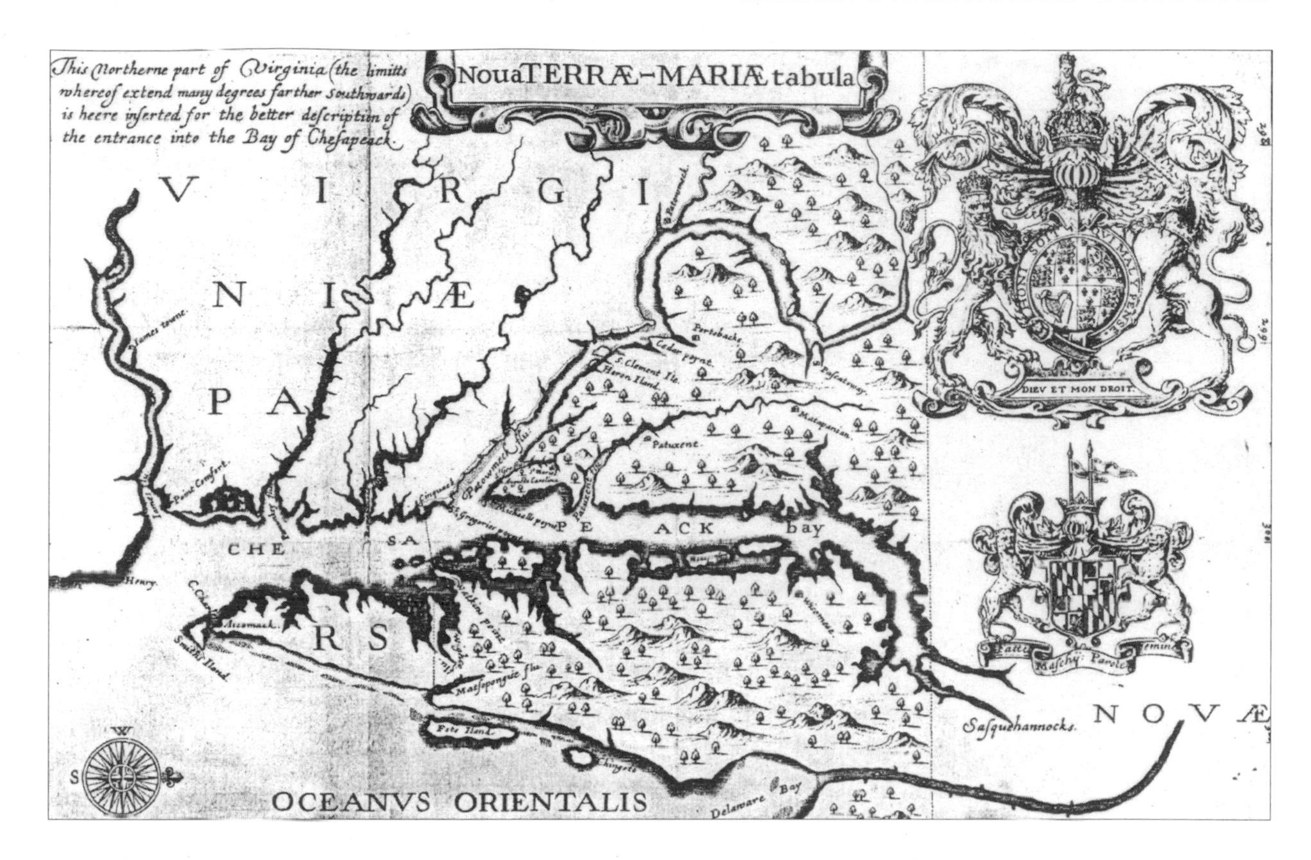

the colony. He thought English Catholics could live in harmony there with Protestants.

Many Catholics did go to Maryland, but not as many as expected. Even Cecil, the new Lord Baltimore, didn't go—and he owned the place. Cecil just stayed home and took the money that came from his colony.

He sent his younger brother, Leonard, to be the first governor. He told him to be "very careful to preserve unity and peace…and treat

Above is Lord Baltimore's 1635 map of *Terra Maria*—which means Maryland in Latin. *Oceanus Orientalis* is Latin for "eastern ocean"—the Atlantic. A century later, Baltimore (below) was a small but thriving township.

At top left is George Calvert, who sought freedom for Catholics in Maryland. Below is his son Cecil, the colony's founder: at right, painted in 1761, is Cecil's great-great-great-great-grandson, Charles—and, of course, a slave.

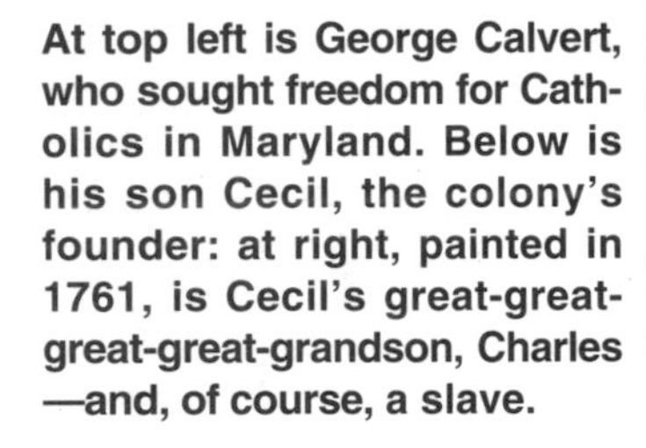

the Protestants with as much mildness and favor as justice will permit."

The Calverts ruled well. They saw that there was representative government and that people were treated fairly. The Calverts urged the Maryland Assembly to pass a Toleration Act. The Assembly did, in 1649. It was meant to protect the Catholic minority and to encourage settlement. That Toleration Act allowed for free exercise of religion—but only for Christians.

Anyone who did not believe in Christ was to be hanged. That meant Jews or atheists (people who don't believe there is a God), or even some Christians who asked too many questions. Those people had to keep quiet or leave Maryland.

It was also a hanging offense to curse God. If you made fun of Christian doctrine, the law said you were to be whipped in public.

That sounds harsh—and it was—but remember the times. In the colonies controlled by France, Spain, and Portugal no Protestants, Jews, or atheists were allowed at all.

Bound Out

A big problem for early colonists in America was the shortage of labor. There weren't enough people to farm and do all the other work. One way of dealing with this was slavery—and Maryland imported many slaves. Another was by bringing over servants and apprentices who were "bound out" to an employer for a certain number of years–four or five at least, and, if you were a boy or girl, often until you were 21. The employer fed, housed, and clothed you. In return, you had to work unpaid, usually learning a trade. If you had a good employer it probably wasn't a bad life; but if your employer was stingy or cruel, or just indifferent, it could be miserable. The newspapers were always full of wanted notices from employers whose bound men or apprentices had run away.

In Maryland, if you stuck the job out, the law said that when your term of service was up you were to be given 50 acres of land, a suit of clothes, an ax, two hoes, and three barrels of corn. Then, if you were a Christian white man, you could vote and be elected to the assembly.

A glassblower and his apprentice.

31 Carry Me Back to Ole Virginny

Sir William Berkeley, twice governor of Virginia in the 17th century, was a thoroughgoing Anglican and a royalist, or supporter of the English monarchy. He thanked God that there were no free schools or printing presses in Virginia in his time, "for learning has brought disobedience and heresy and sects into the world; and printing has divulged them and libels against the government. God keep us from both."

On this optimistic English tobacco label, colonists and an Indian chat happily together about the merits of the brand being advertised.

I haven't said much about Virginia since John Smith's time, and that's too bad, because some future presidents were getting born there: George Washington, Thomas Jefferson, James Madison, and James Monroe. Each became president. Virginia's John Marshall, George Mason, and Patrick Henry were great leaders, too.

Was there something about Virginia that bred leaders? Life there was certainly different from life in Pennsylvania or Massachusetts.

Most people in 18th-century Virginia weren't Puritans or Quakers. They were Anglicans: members of the Church of England. But they were more relaxed about their religion than the Puritans. In the New England colonies the ministers were the most important people in the community; in the South the wealthy landowners were more important.

The Virginians didn't live in towns, as people did in Massachusetts. They lived along the rivers on small farms, or on very large farms called plantations. Living on the river made shipping easy, and that was important.

What Virginians were shipping was tobacco—to England. While a few other crops were grown, the main

The slaves on a tobacco plantation don't just plant and pick and dry and cure the tobacco. The slaves do *all* the physical labor—here they are making the barrels to ship the tobacco, and packing them, too.

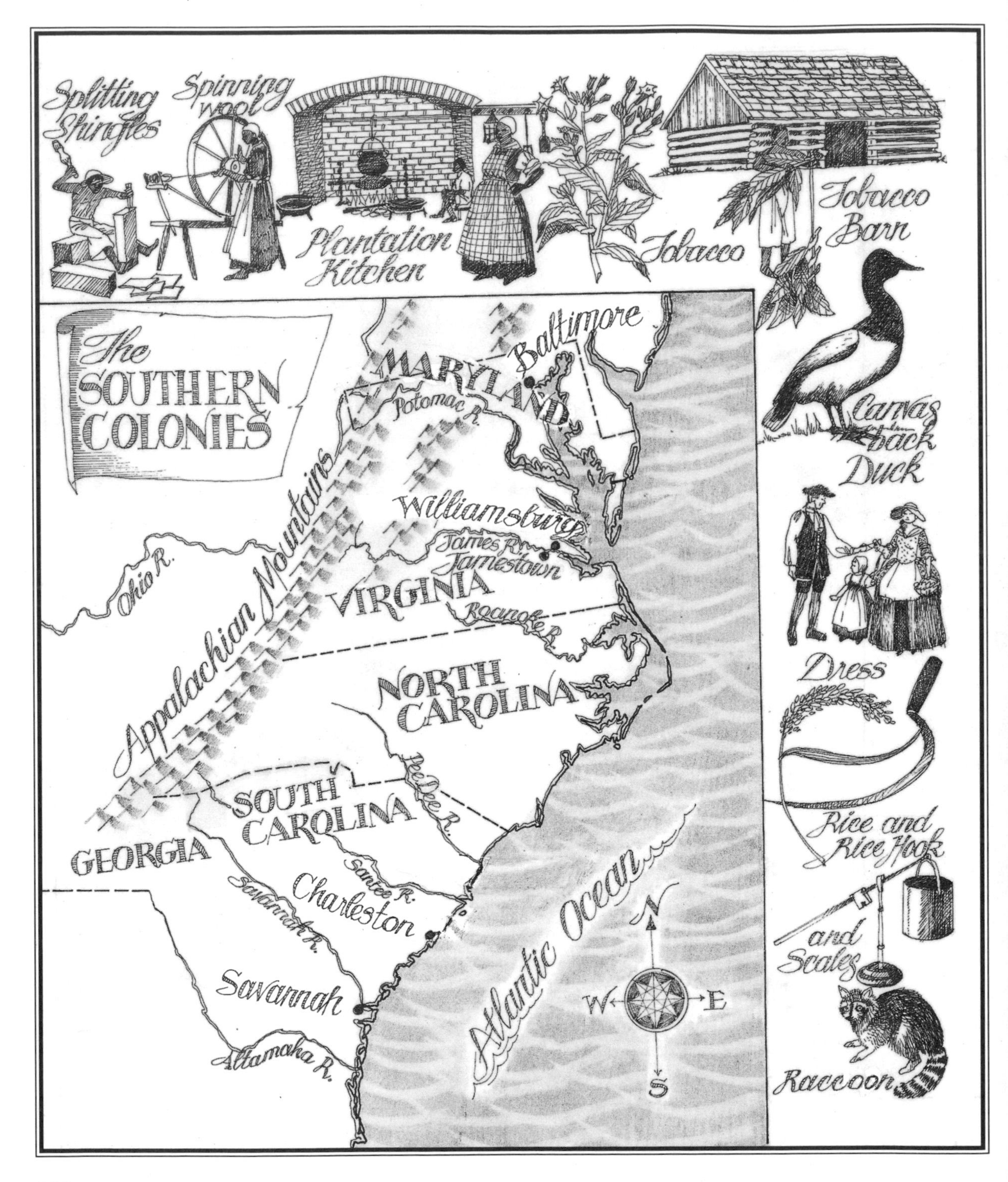
Splitting Shingles
Spinning wool
Plantation Kitchen
Tobacco
Tobacco Barn
The SOUTHERN COLONIES
Baltimore
MARYLAND
Potomac R.
Canvas back Duck
Williamsburg
James R.
Jamestown
Ohio R.
Appalachian Mountains
VIRGINIA
Roanoke R.
Dress
NORTH CAROLINA
Pee Dee R.
SOUTH CAROLINA
Rice and Rice Hook
GEORGIA
Santee R.
Savannah R.
Charleston
Atlantic Ocean
N
W
E
S
and Scales
Savannah
Altamaha R.
Raccoon

money-maker was tobacco. That was a problem. When tobacco prices were high, Virginians did well. When tobacco prices fell, they were in trouble. There was no balance.

There was little industry, and most goods came from England. In this land of magnificent forests, even fine furniture and other wood products were shipped from the "mother country."

There was another problem. Tobacco uses up the soil. After a few years, nothing grows well on land that has been planted with tobacco. To succeed in a tobacco economy, you need to rest the land every few years. That means you need to own a lot of land. You also need many workers.

So tobacco growers in colonial Virginia began to buy land and workers. As you know, at first they bought indentured servants. Then they bought Africans and made them slaves. By 1750 there were more Africans in Virginia than any other single group of people. More Virginians had come from Africa than from England or Scotland.

But it was a few rich white planters who held power in the colony. Most whites were small farmers, and there were thousands of them. Some owned a slave or two—or hoped to—and that made them go along with the big slave owners. Virginia was not the only place where this happened. A society built on slavery stretched from Maryland to Georgia. Slavery was not only terrible for the black slaves—it ruined many white farmers, too.

At first the South, like the North, was full of yeoman farmers. Yeoman are independent farmers who work their own land. When slave ships began bringing in large numbers of Africans, the yeomen were in trouble. The blacks, being slaves, were forced to work for nothing. The yeomen farmers couldn't compete with that. The tobacco they grew was more expensive than tobacco grown by slaves.

The yeomen farmers had these choices. They could stay in Virginia (or Maryland or the Carolinas or Georgia) and try and work their own farms. Usually that meant they would become "poor whites."

Or they could buy slaves.

Or they could head West, to the frontier, and settle new land (and perhaps fight Indians for that land). These were difficult choices.

By the 18th century most slave owners were beginning to realize that slavery was wrong. Many spoke out against it. (They also made excuses and tried to justify enslaving others. They knew that every ancient society had included slaves.) Many whites believed they were trapped in a bad system; they didn't know how to get out. George Washington, Thomas Jefferson, and others wrote that slavery was evil—but they owned slaves. If you were a plantation owner and you freed your slaves, you might become poor. So might your family.

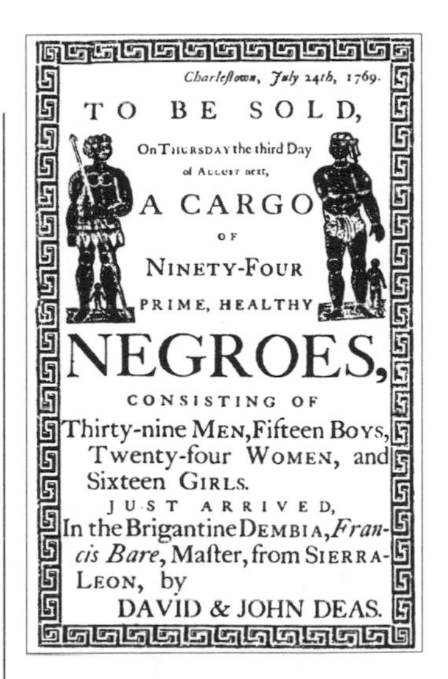

A slave auction notice. Traders usually wanted slaves to look young and healthy—they were worth more. But they didn't like slaves who looked proud and held their heads high. Such people might be troublemakers.

Divorce, Colonial Style

Dame Alice Clawson, who lived on Virginia's eastern shore in the mid-1600s, was a woman who wouldn't take nonsense from anyone. When her husband came home from a stay with the Nanticoke Indians, he brought an Indian woman home with him. Dame Alice was outraged, and hauled her two-timing husband off to the local justices. She became the first Virginia woman to obtain a divorce.

A plantation owner (inset), a mansion, and rows of slave cabins—with dirt floors, no glass in the windows, no water (except a pump everyone shared), and smoky open fireplaces.

What would you do?

For a decent person, the problem was even more difficult than you can imagine. Many Virginians had been very poor back in England or wherever they or their parents came from. Many had been let out of jail if they would agree to come to America. They knew what it was like to be oppressed, and many didn't want anything to do with slavery. But in some places it was against the law to free slaves. In some places it was against the law for black people to own land. It was against the law to teach black people to read. Many slave owners believed that if they freed their slaves, there would be no way for them to survive. Most would not be able to find jobs. They might starve. They might be kidnapped and sold again into slavery. A few whites did free their slaves. Some tried to end slavery. Most did nothing. What would you have done?

Imagine that you are a slave owner. Imagine that you control other people's lives. Imagine that you can order people around. Imagine that you can have them whipped if they don't do what you tell them. Do you think that might make you lazy? Do you think you might turn into an angry tyrant? Do you think you might become mean?

Some slave owners were all of those things. Some of them were among the meanest people in America's history. But there were others who felt great responsibility for their slaves.

There is a paradox connected with slavery. A paradox is a puzzle. Something very puzzling happened in Colonial Virginia.

When it came time to write a constitution for our nation, it was the slave-owning Virginians who thought and wrote most about freedom. That is the paradox. Why do you think it was so?

After seeing some slaves try to rebel against their owner, an observer from New Jersey wrote: "The ill Treatment which this unhappy part of mankind receives here, would almost justify them in any desperate attempt for gaining that *Civilitie*, & *Plenty* which tho' denied them, is here commonly bestowed on Horses!"

32 The Good Life

For most of the 18th century wigs for the rich got fancier and fancier.

We haven't finished with paradoxical Virginia. We still haven't figured out why so many great leaders were growing up in that colony. Could it be that, for the very few lucky enough to be born wealthy, there was time to become well educated and time to be spent thinking?

By the 18th century, a plantation owner lived a privileged, lordly life. But it wasn't as easy as some people think. To run a plantation well, you needed to be intelligent and industrious. (In England, aristocrats often didn't work—or want to work; that wasn't true in America.)

Each plantation was like a small village owned by one family. That family lived in a great house with many rooms and many servants. The house was usually built of brick and had a long lawn that led to the river. The kitchen was a separate small building. So was the laundry, the carpenter's shop, the spinning and weaving shed, the blacksmith's shed, and the plantation office.

The plantation's business was farming, so stables and barns and buildings for the farm's produce were needed. There was a smokehouse for smoking meats. (Smoking preserves meat, and in those days there were no refrigerators.) Of course there was a dock for shipping tobacco. Tobacco was packed into barrels, called hogsheads, so there was a shed where the

An English visitor wrote home in 1759: "Solomon in all his Glory was not array'd like one of these [Virginians]. I assure you, [even] the common Planter's Daughters here go every Day in finer Cloaths than I have seen content you for a Summer's Sunday. You thought...my Sattin Wastecoat was a fine best...I'm nothing amongst the Lace and Lac'd fellows...here."

This little girl lived in Maryland. She was six years old when her picture was painted in about 1710. Her clothes would be fun to dress up in—not for climbing trees, though.

Dancing was always part of an upper-class Southern education. In the 18th century, with no stereos or TV, people had to make their own entertainment.

cooper, or barrelmaker, worked. The plantation even had a kiln, where bricks were baked.

Slaves lived in cabins built near the fields. On the big plantations there were sometimes 200 or more slaves. A man known as "King" Carter, the wealthiest Virginian, owned 10 plantations. Some other Virginians owned two or three plantations. All the people who lived on the plantation needed to eat, so vegetables and corn and wheat were grown, and animals were raised.

Can you see why it was hard work to run a plantation well?

A plantation owner was like a business executive. He ran the plantation and saw that it made a profit. He organized and fed and clothed all those people. He was probably a member of the House of Burgesses, and that meant that he attended assembly sessions at the capital twice a year.

He may also have been a court officer, called on to decide court cases. Besides, he and his wife gave parties and entertained visitors, who sometimes stayed for days and days or even weeks at a time. It was a busy life.

Plantation children didn't live at all the way you do. Some of the ways they lived were nice, but some you wouldn't like. We know you wouldn't like it if you were a very rich planter's son and had to wear velvet pants and ruffled shirts and high-heeled shoes, just like your dad. We don't think you'd like to shave your head so you could wear a powdered wig. When you wanted to play you wore an embroidered cap instead of that powdered wig. (Only the very rich went in for head shaving.)

If you were a planter's daughter, you would wear satin gowns with stiff petticoats. Now that

Slave or Bondman?

William Byrd was a wealthy plantation owner who compared himself to men of the Bible—the patriarchs (PAY-tree-arks). That way, he made himself believe slavery was all right. He never used the word "slave." He said "bondman" and "bondwoman." He was using a euphemism (YOO-fuh-miz-um): a "nice" word that you use because the real word is embarrassing. (Do you know some euphemisms that we use today?) Byrd thought he lived an ideal life, and he wrote about it in his diary:

I have a large family of my own, and my Doors are open to Every Body, yet I have no Bills to pay, and half-a-Crown will rest undisturbed in my Pocket for many Moons together. Like one of the Patriarchs, I have my Flocks and my Herds, my Bond-men and Bond-women, and every Sort of trade amongst my own Servants, so that I live in a kind of Independence from everyone but Providence.

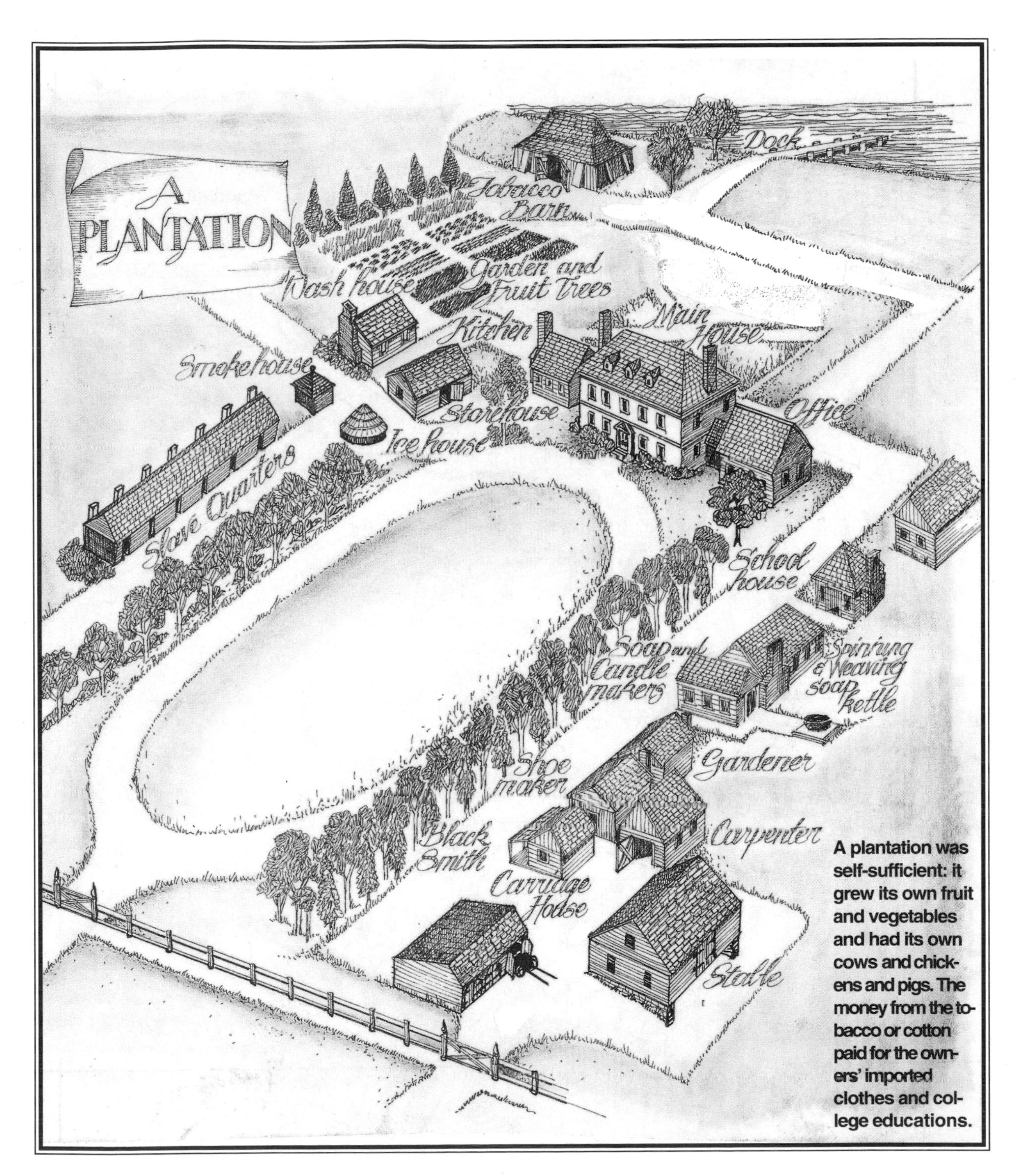

A plantation was self-sufficient: it grew its own fruit and vegetables and had its own cows and chickens and pigs. The money from the tobacco or cotton paid for the owners' imported clothes and college educations.

Bedtime Stories

Do your parents sometimes tell you to *sleep tight*? In colonial times mattresses were laid over ropes stretched between the sides of the bed frame. When the ropes sagged, they had to be tightened with a wind-up wrench. So people really meant it when they talked about sleeping tight.

How about *climbing into bed*? Do you use that expression? And do you climb into bed, or is your bed low enough so you can just fall into it? In colonial days, there was no central heating and the floor got very cold at night. So beds were tall, and most people had to actually climb a few steps to get to bed.

might be fun, because those were party clothes. Everyday clothing for boys and girls was more comfortable, but not like your jeans and shorts.

One thing you would like in colonial times was horseback riding. Everyone learned to ride horses, and everyone learned to dance. Dancing was very popular.

Pretend you are the child of plantation owners: you have your own schoolteacher who lives with you.

Life is good to you. You study and play and go to parties. You eat big meals of meats and pies and vegetables, all home-cooked. Slaves pick up after you. How do you treat them? Some boys and girls are considerate; some are not.

If you are smart and study hard, you will be taught to read the Bible in the languages in which it was written. You will learn about ancient Rome and its gladiators, poets, and politicians. You will play a musical instrument. Many Americans born after you will not have as rigorous an education.

If you are a boy you may finish your schooling at Virginia's College of William and Mary or perhaps at college in England. Someday you will be expected to serve in the House of Burgesses. You are being trained to be a leader.

The College of William and Mary in Williamsburg was named for England's king and queen and was the second college in the colonies.

Even if she hated it, every little girl had to learn to sew and embroider. Girls practiced on samplers like this one—it includes 12 different kinds of stitches.

33 Virginia's Capital

Virginia's enterprising Governor Alexander Spotswood started an iron industry and explored the Shenandoah valley.

Do you remember the mosquitoes, deerflies, and snakes at Jamestown, back in John Smith's time? Well, things got better on that swampy peninsula—but not a lot better. In 1676 some frontiersmen became angry and marched to Jamestown. They wanted the colonial government to help them fight Indians, and they hated the governor, Sir William Berkeley (BARK-lee), who was a tyrant. Even the king called him "an old fool." The rebels burned the State House.

Almost as soon as the State House was rebuilt, it burned down again. This time it may have been an accident. Still, in 1699, when someone suggested moving the Virginia capital to Middle Plantation—eight miles away but on higher ground—most people thought it a fine idea.

Actually there were at least two "someones" who did the suggesting. One was James Blair, a Scotsman who was minister of Bruton Parish Church, the Anglican church at Middle Plantation. Reverend Blair was also the founder and first president of the College of William and Mary, which was at Middle Plantation. The other someone was Francis Nicholson, the new royal governor. Nicholson agreed that with a church and college in place, Middle Plantation would be a capital spot for the capital. But Middle Plantation wasn't much of a name for a town. A new name was needed. The King of England's name was a natural choice, and the town became Williamsburg.

That little town, born at the beginning of the 18th century, danced across the stage of history for about 80 years. Then it left the spotlight and was forgotten (until the 20th century, when it was restored and rebuilt as if it were still in the 1700s).

Today we think the rebels—led by Nathaniel Bacon—were wrong in their attitude toward the Indians, but right in their feelings about Governor Berkeley. The Virginia leaders knew about King Philip's War in New England, and they didn't want to start anything with Indians.

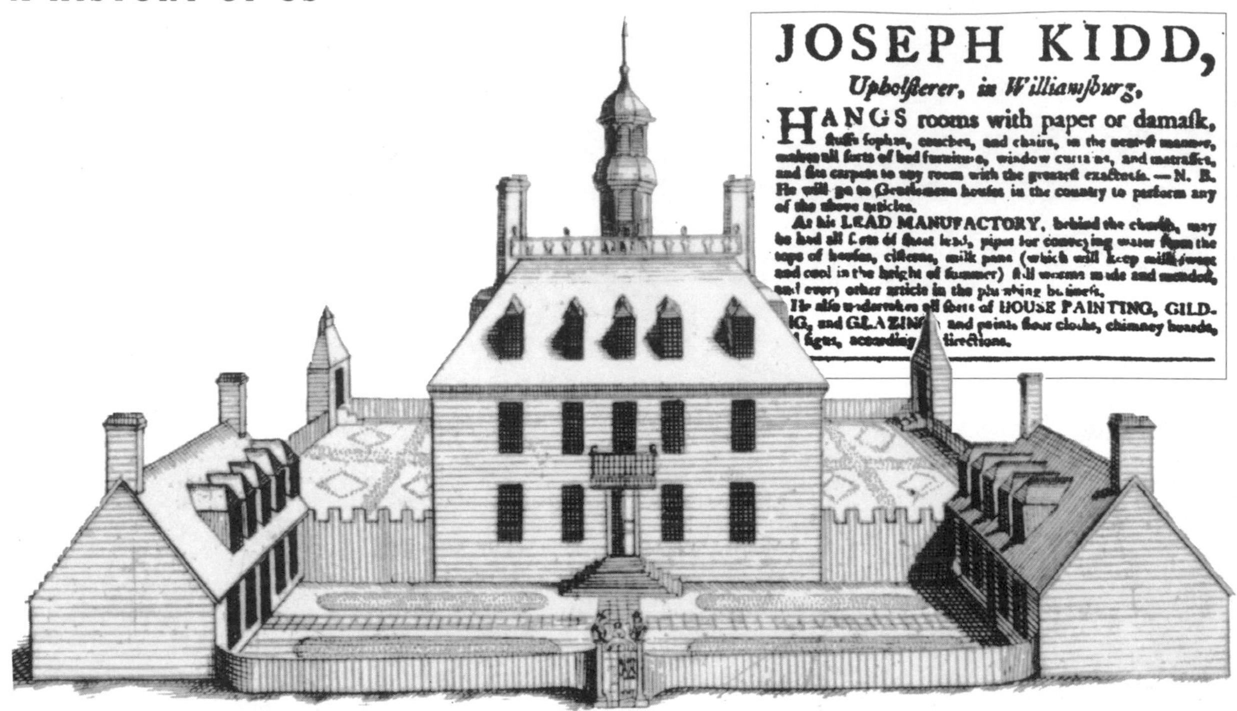

Lord Botetourt, a popular English governor, was the first to fully occupy the Williamsburg governor's palace (above). His groom of the chambers, Joseph Kidd, started his own decorating business.

Francis Nicholson arrived in Williamsburg fresh from Maryland, where he had been governor and had designed Maryland's capital, the bustling port of Annapolis. He was eager to plan Williamsburg, and he did it with care, using new ideas from France.

Williamsburg was "laid out regularly in lots...sufficient each for a house and garden and...free passage for the air which is grateful in violent hot weather." Wide, tree-lined Duke of Gloucester Street ran down the middle of the town, with the Capitol at one end and the college's Wren Building at the other end. Bruton Parish Church, with its white spire, stood proudly on the same street, along with neat houses, shops, and a grassy mall. Overlooking the mall was the finest building in the colony: the handsome, stately governor's palace. Now picture people, horses, cows, sheep, and gardens, and you have an idea of Williamsburg.

The word ***crown***, as it is used here, means the British government.

If you stand outside the governor's palace and look at the brick fence—with its stone British unicorn on top and fancy iron gates—you will be impressed. If you are lucky you may get invited inside for a musical evening. You'll sit in the candlelit ballroom, where men in starched linen blouses and women in silk brocade gowns smile and nod at each other. A display of muskets and swords in the entry is intended to leave you awed with the Crown's military might.

Jamestown's landowners didn't think moving the capital was a good idea, but most of the other plantation owners did.

But, best of all, most young people agree, is a chance to lose yourself in the boxwood maze garden growing behind the palace. Almost everyone knows the ancient myth of the maze and the monstrous

Minotaur on the island of Crete. The maze behind the governor's palace doesn't have any monsters—there are some peacocks strutting around—but people like to go in it and pretend they are lost, especially if there are two of them and they are in love.

Most of the year, Williamsburg is a sleepy village of 2,000 souls. Half of them are African-Americans. But in April and October, when the House of Burgesses and the courts are in session, the population doubles. Men sleep three to a bed at Christiana Campbell's tavern, where the good Mrs. Campbell, "a little old Woman, about four feet high & equally thick," keeps a popular dining room.

These are Public Times, and people come to take care of colonial business, make laws, consider court cases, see friends, and shop. You can buy a wig at the wigmakers, a violin at the music shop, or a gingerbread cookie at the bakery. If you want "to put on the dog," you can have doghide shoes made for you at the bootmakers. You might go to an auction, a horse race, or a fair. You can see a play. (The first theater in the English colonies opens in 1717.)

One traveler writes of Williamsburg, "At the time of the assemblies and general courts, it is crowded with the gentry of the country. On those occasions there are balls and other amusements; but as soon as business is finished, they return to

Joy and Loyalty

In July 1746 news of the English army's crushing victory over rebel Scots at Culloden reached Williamsburg, and was celebrated in style, as reported in the Virginia Gazette*:*

In the Evening, a very numerous Company of Gentlemen and Ladies...after dancing some Time, withdrew to Supper...consisting of near 100 Dishes...There was also provided a great Variety of the choicest Liquors, in which the Healths of the King...and the rest of the Royal Family, the Governor, Success to His Majesty's Arms, Prosperity to this Colony, and many other Loyal Healths were chearfully drank, and a Round of the Cannon...was discharg'd at each Health, to the Number of 18 or 20 Rounds...

All the Houses in the City were illuminated, and a very large Bonfire was made in the Market-Place, 3 Hogsheads of Punch given to the Populace; and the whole concluded with the greatest Demonstrations of Joy and Loyalty.

The Rev. Blair didn't like fun and games, but he took charge of William and Mary and made it a good place to go to college.

Cooks at work in a big 18th-century kitchen like that of the governor's mansion. Dinners were huge and often consisted of a dozen courses or more—with fish, beef, and turkey all in the same meal.

their plantations and the town is in a manner deserted."

For most people a trip to Williamsburg means having a good time—unless the Reverend Mr. Blair happens to come by. Blair doesn't have a sense of humor. For fifty long years, he is a powerful force in the community. It is said that three governors have been recalled to London because of his complaints. A fourth governor remembered him as "a very vile old fellow." Blair usually gets his way. And his way is not one of tolerance. A landowning aristocracy rules Virginia and outsiders remain outside. Only Anglicans can be elected burgesses. A plan for religious liberty is hatched in Williamsburg in the second half of the 18th century, but it might not have happened if Reverend Blair had been alive.

But Blair is mortal and not typical of most Virginians. The Virginia way is gracious and courtly and easygoing. William and Mary's wise law professor, George Wythe (say *with*), will soon come to prominence in Williamsburg. He is kind and considerate and learned; some people call him

School's Out

Virginia's governor Alexander Spotswood wanted to be friends with the Indians. He sponsored the Indian Act of 1714. It created an outpost where the colonists could meet and trade with the Indians. An English school was founded to teach Indian children about Christianity and English ways.

Many colonists were angry about Spotswood's ideas. Some didn't want anything to do with the Native Americans. Others were afraid the Indians would sell their furs to Virginia's governor and not to them. In 1717, at the colonists urging, the king repealed the Indian Act. The outpost and the Indian school were closed.

"a walking library." George Wythe becomes mayor of Williamsburg. He frees all his slaves long before he dies. He hates slavery. One of his favorite students is a young man named Thomas Jefferson.

If you had a magic wand that could waft you anywhere you wanted, you might consider landing at the Raleigh Tavern in Williamsburg on a day when George Washington, Thomas Jefferson, Patrick Henry, and Peyton Randolph are having a conversation. Just sit back and listen. Those Virginians and their friends have a rare talent for good times and serious thought. They are as splendid and energetic a group of leaders as any nation has ever produced.

They are struggling with a serious problem: how to create a new and fair government on this splendid continent. And they are being pulled in two directions. There are many who think as James Blair does and want to reproduce Old England in America. One professor at the College of William and Mary writes approvingly that people in Williamsburg "behave themselves exactly like the gentry in London."

But there are others who don't behave like English gentry. They have come to America because they are unhappy with the Old World and its society of rigid classes. They want to try something new, in what they call the New World. This little village of Williamsburg will ring with debates on the purposes of government. Should individual liberty be the goal? Or is it better to preserve an orderly community where everyone has a sure place? What does "the consent of the governed" really mean? Wouldn't you like to hear what Thomas Jefferson and George Washington have to say about those questions?

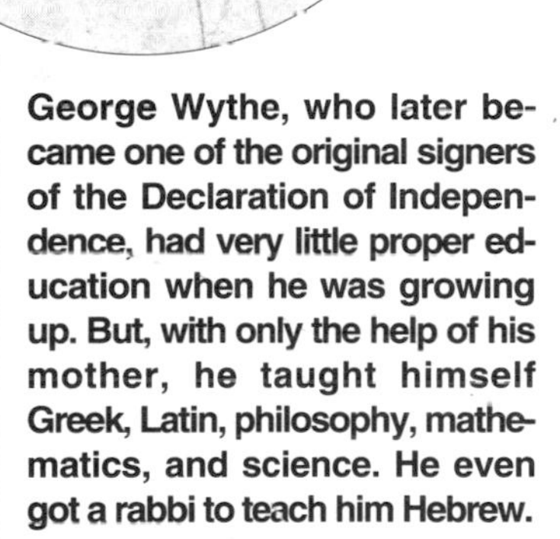

George Wythe, who later became one of the original signers of the Declaration of Independence, had very little proper education when he was growing up. But, with only the help of his mother, he taught himself Greek, Latin, philosophy, mathematics, and science. He even got a rabbi to teach him Hebrew.

Colonists resisted Spotswood when he wanted an Indian school. But in 1723 this building was completed, and a small group of Indians—a dozen at a time—lived and studied there.

34 Pretend Some More

Unliberated

Were women treated equally in the colonies? See what you think. A woman sent this poem to the *Virginia Gazette*. It was published in October 1736.

Custom alas! doth partial prove,
Nor give us equal Measure;
A Pain for us it is to love,
But is to Men a Pleasure.

They plainly can their thoughts disclose,
While ours must burn within;
We have got Tongues, and Eyes, in Vain,
And Truth from us is Sin.

Then Equal Laws let Custom find,
And neither Sex oppress,
More Freedom give to Womankind,
Or give to Mankind less.

Venerate means "respect." A farmer's plow was the most important machine he owned.

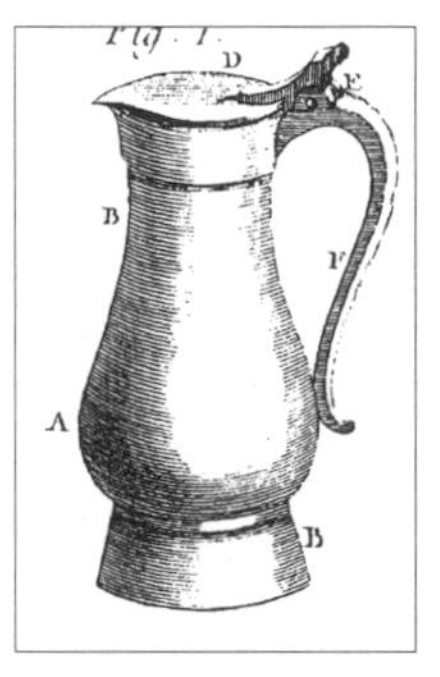

Silversmiths also made pewter items, like this tankard with its flip-up lid.

You are 10 years old and indentured to Patrick Beech, a silversmith in Williamsburg. Beech was a real silversmith, we know that, but we don't know much about his servants and apprentices. We do know about the jobs they would have done and the way they might have lived. So you can pretend to be one of them. If you do, you will get an idea of what life was like for some children in the 17th century.

You have been in Virginia just a year. Your parents died of influenza in London. Since you had nowhere to go and no money, the Lord Mayor sent you and some other orphans to Virginia. Sometimes you are homesick for your friends in London, but you are beginning to like it here in America. Life isn't easy, but it is better than it was in London.

Beech keeps you working from before the sun comes up until dark. Your first job in the morning is to light a fire to warm the house and another to heat the forge where silver is melted and formed. Then you clean the kitchen, run errands, and sometimes do odd jobs in the silver shop. You don't get time to play, and lately you have begun to hate Mr. Beech. He never seems to smile.

Mrs. Beech has taught you to read. She takes you to church and whips you when you are bad. She does the same with her children. She is fair but very busy. You miss your own mother.

But you do have a friend in the silver shop. His name is Tom, and he was apprenticed to a watchmaker in London before he came to the colonies. Tom says Beech is not a bad man, just worried. There isn't enough work in Williamsburg. There are too many silversmiths, and the rich planters buy their good silver in England. So Beech must make silver teeth and set them in people's mouths. He repairs watches and makes clocks, and sometimes silver cups and trays and jewelry. He has

a big family to feed, along with his servants. There is enough food, but nothing extra.

Neither you nor Tom is free. The law says you must stay with Patrick Beech until you are 21. Tom is also an indentured servant. Beech paid for his passage from London: he must work for him for five years. Then, Tom tells you, he will go to Charles Town and open his own shop. He has heard that the planters in South Carolina are very rich.

More than anything, you want to hunt and fish and learn to use a rifle. You have seen enough of the silversmith's work—fires and forges, pouring liquid metal, hammering silver into shape, putting teeth into the jaws of people who scream in pain. You need to be patient. Someday you will have everything you wish for—and more. You have energy and ambition and you will become a farmer.

Now, pretend again. This time you are the child of a Virginia farmer. You live in a small wooden house with a big fireplace at one end. The house has only one big room with a sleeping loft. At night your parents sleep in front of the fire with the baby, and you and your brothers climb a ladder to the loft. You all sleep on straw mattresses.

The only clothes you have are those you wear, and a Sunday shirt. Tobacco and corn grow poorly on your land, because it is worn out. You have enough food, but you don't eat a balanced diet. You are sometimes sick, and you will die before you are 40.

Still, you are luckier than many children. You have had a year of schooling, and you know how to read. Your parents hope your life will be better than theirs. And it will be. At 15 you will head out to the western frontier, where you will find land and opportunity. Your schooling will help you succeed. You will marry, have 10 children, and own land enough for all of them to farm.

Most poor white boys and girls don't get to school. And there are no schools for black children. Many 18th-century Southern children never learn to read or write. It is difficult to have schools when people live so far apart on farms. It is difficult to have churches, too. Some ministers ride horseback from one church to another. The law says you must go

Forty Shillings Reward.

RAN away this morning from the ſubſcriber, an Apprentice BOY named James Hoy, near 18 years old, about 5 feet 8 inches high, fair complexion, and hair tied behind. He took with him a brown jean coat, grey ſurtout, new ſhoes, round felt hat, and many other articles of cloathing.——Whoever brings him back or confines him in gaol, ſo that he be had again, ſhall receive the above reward.

JOHN FARRAN.

Philadelphia, June 26. 3ſp

This is a pewter works. Pewter, which is a mixture of tin, lead, and copper or bismuth, was for people who couldn't afford silver. The apprentice is turning the wheel that drives the bellows to keep the furnace hot. If the 'prentice got fed up and ran away, his master would offer a reward for bringing him back. In the colonies, even unskilled laborers were hard to find.

No one knows who painted this picture, but it may have been done in South Carolina. The slaves may be celebrating a wedding by "jumping the broomstick." Some of their clothes look like those worn in Yoruba, which is in Nigeria in Africa. The banjo-like instrument (right) resembles a Yoruban *molo*; the other instrument looks like an African *gudu-gudu*.

to the Anglican church every Sunday.

Your parents hate that law, and another law that makes them pay taxes to help support the Anglican church. They say the Anglican church is for rich folks. They would like to join the Baptist church, but they can get in trouble if they do that. A law says their children could be taken from them if they join a freethinking church. Since you are one of their children you agree that is a terrible law. Actually, it isn't enforced very often. Perhaps the law is meant to scare people away from new churches: like those of the Baptists and Methodists and New Light Presbyterians.

Belonging to that Anglican Church of England, as most Virginians do, makes you different from people in Massachusetts. Virginians love England and English clothing, paintings, furnishings, and ideas. They feel closer to the people in England than they do to those in New England.

Now pretend that you are a slave. You don't want to? You are right. No one wants to be a slave. Some slaves, especially those who work in the fields on some big plantations, live in small huts and sleep on old blankets piled on the dirt floor. They don't eat well and they work almost all the time. Other slaves, especially those who are house servants, live in small wooden houses with beds and tables and furnishings that come from the plantation workshop or, sometimes, from the big house. Visitors from Europe will say they live better than most peasants in the Old World. (Which is probably why so many European peasants want to come to America.)

Your name is Sarah, and you live on a farm in North Carolina. There are 16 of you slaves (including children) in four families and you share two houses near the tobacco fields. A fireplace divides each house in half. Your house has a front porch, as many houses do in Africa. The porch is a nice place to rest on a hot evening. You have your own garden, which provides corn and greens and potatoes. They have made you strong and healthy. You are 11, and you can't read and never will be taught how. But you can sing and play the banjo, and that makes you popular in church. Your faith is important to you and your family. You are Christians now, and you have brought your African spirituality to that religion.

Hoeing Cotton

From the story of a former slave, Solomon Northup:

About the first of July, when [the cotton] is a foot high or thereabouts, it is hoed the fourth and last time...During all these hoeings the overseer or driver follows the slaves on horseback with a whip... The fastest hoer takes the lead row. He is usually about a rod in front of his companions. If one of them passes him, he is whipped. If one falls behind or is a moment idle, he is whipped. In fact, the lash is flying from morning until night.

35 Carolina: Riches, Rice, Slaves

An old saying goes: "Carolina in the spring is a paradise, in the summer a hell, and in the autumn a hospital." All the rich planters left their steamy plantations in May and came to Charlestown for six months to escape yellow fever and malaria—and to have some fun in the big city.

The seal of the Carolinas viewed Indians more kindly than most colonists did.

The Carolinas, North and South, were granted to eight lords proprietors by King Charles II. The lords never meant to live in America, and they didn't. They just planned to get rich by using the Carolinas to produce three products that were expensive in England: wine, silk, and olive oil.

The Carolinas worked out, but not as the lords had expected. Indigo, a plant grown for its blue dye, and rice became principal crops—not wine, silk, or olive oil. Eventually the colonies were bought back by the king. The lords lost out, and the Carolinas became royal colonies.

In South Carolina, Charles Town, named for Charles II, prospered from its beginning. No longer were there terrible starving and dying times when a colony was founded. Jamestown and Plymouth had taught the colonists what not to do.

Charleston (which was the name Charles Town turned into) soon became the busiest port in the South. It attracted younger sons of the English nobility, who gave it an aristocratic flavor.

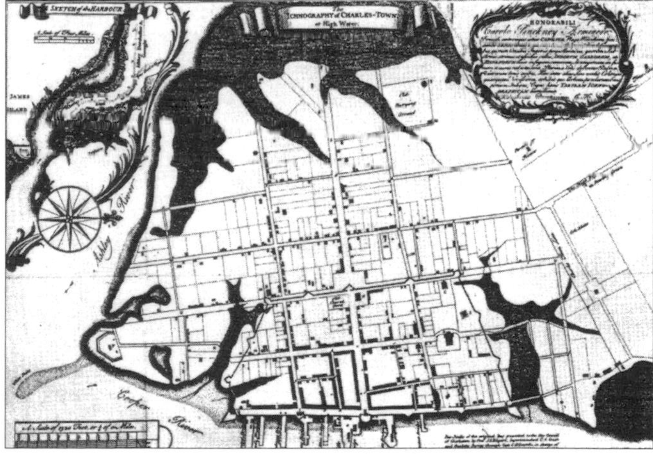

Unlike Boston or New Amsterdam, Charleston's and Philadelphia's streets were planned before any houses were built.

Many of Charleston's leaders came from the island of Barbados in the Caribbean Sea. In the 17th century Barbados was the wealthiest and most crowded of all the colonies in English America. (In 1680 the exports of that tiny island were more

Here Peter Manigault, a well-known Charleston Huguenot, seems to be drinking too much. (He's the one holding a bottle on the left.) Drinking was sometimes a problem in early America. (Is it today?)

One Man's View

In 1769 "Capt. Martin a Man of War" wrote this poem about Charleston:

Black and white all mix'd together,
Inconstant, strange, unhealthful weather,
Burning heat and chilling cold
Dangerous both to young and old
Boisterous winds and heavy rains
Fevers and rheumatic pains
Agues plenty without doubt
Sores, boils, the prickling heat and gout
Musquitos on the skin make blotches
Centipedes and large cockroaches
Frightful creatures in the waters
Porpoises, sharks and alligators
Houses built on barren land
No lamps or lights, but streets of sand
Pleasant walks, if you can find 'em
Scandalous tongues, if any mind 'em
The markets dear and little money
Large potatoes, sweet as honey
Water bad, past all drinking
Men and women without thinking
Every thing at a high price
But rum, hominy and rice
Many a widow not unwilling
Many a beau not worth a shilling
Many a bargain, if you strike it
This is Charles-town, how do you like it.

valuable than the exports of all of the North American mainland.)

South Carolina practiced religious tolerance and that led an interesting mixture of settlers to the colony. Scots settled on the coast and helped fight off Spanish attacks. French Protestants, called Huguenots (HUE-guh-noes), came and became prominent citizens—just the kind of colonists the new land needed. In France they were persecuted, but France's shortsightedness was America's good fortune. The Huguenots were carpenters and blacksmiths and masons, and they believed in hard work. What John Smith would have given to have had them in Jamestown!

Fieldworkers were needed to plant and harvest the rice that was making the colony rich. The settlers from Barbados were used to owning slaves; they wanted slaves in America, and they encouraged it. It is an irony that Africans probably taught the white settlers how to cultivate rice. Rice was grown in Asia and Africa, not Europe. It was rice that made slavery profitable in South Carolina. Soon more black people in South Carolina than whites.

South Carolina became an aristocratic colony with a few very wealthy people holding almost all the economic and political power. It was different from the other

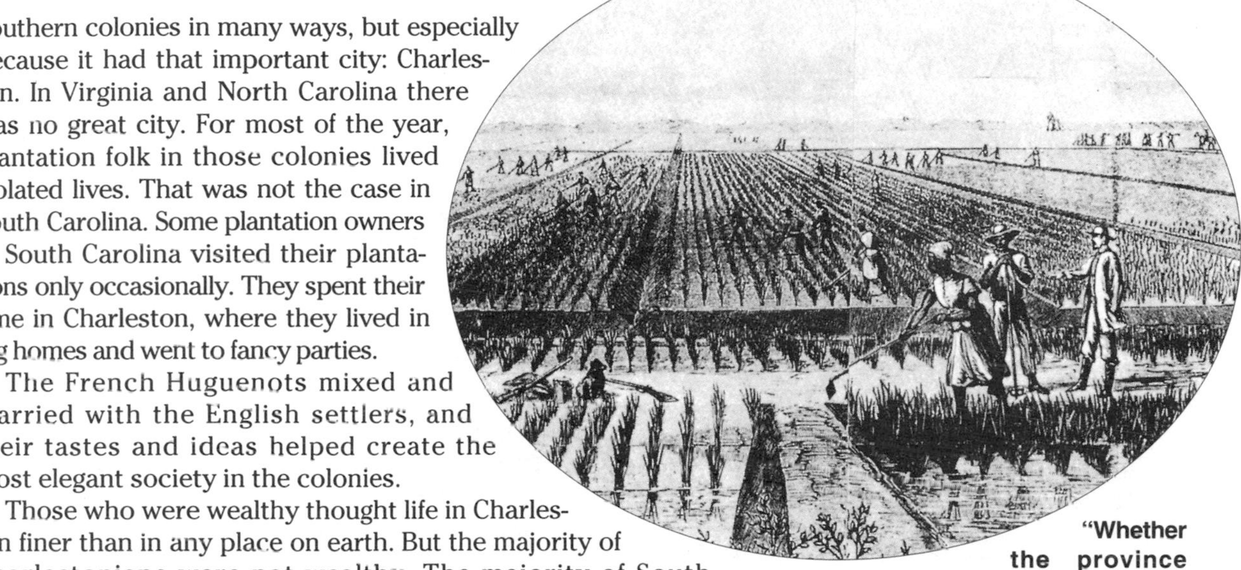

Southern colonies in many ways, but especially because it had that important city: Charleston. In Virginia and North Carolina there was no great city. For most of the year, plantation folk in those colonies lived isolated lives. That was not the case in South Carolina. Some plantation owners in South Carolina visited their plantations only occasionally. They spent their time in Charleston, where they lived in big homes and went to fancy parties.

The French Huguenots mixed and married with the English settlers, and their tastes and ideas helped create the most elegant society in the colonies.

Those who were wealthy thought life in Charleston finer than in any place on earth. But the majority of Charlestonians were not wealthy. The majority of South Carolinians were not free. They were the Africans, and they worked as field hands, craftspeople, and servants. But they had their own ideas and traditions, and they brought those African ideas, songs, stories, and habits to their new home. After a while, some of those African ideas became mixed with ideas they found in America. Brer Rabbit, who started as an African, became American. In South Carolina African-Americans developed their own language, called Gullah.

Some people in South Carolina speak Gullah today. They talk quickly, without a Southern accent. Gullah combines words from English, French, and a number of African languages. You may have heard some Gullah words, such as *goober* (peanut), *gumbo* (soup with okra), *juke* (as in jukebox), and *voodoo* (witchcraft). Here is a sentence in Gullah: *Shishuh tall pass una*. It means "Sister is taller than you."

"Whether the province may have acquired it [rice] by shipwreck, or whether it may have been carried there with slaves, or whether it be sent from England, it is certain that the soil is favorable to it."

In 1775, Charleston was the fourth biggest city in the colonies, with a population of 12,000. The biggest of all was Philadelphia, with 40,000 people. It was bigger than any English city except London.

These are indigo farm workers. Indigo was an ideal second crop because it needs no work in the winter months, which is when all the hard labor of rice planting must be done. The blue dye made from the indigo plant was much prized.

36 Carolina: Dissenters and Pirates

William Byrd *didn't think much of his neighbors in North Carolina. This is what he wrote about them:* "Both Cattle and Hogs ramble in the Neighbouring Marshes and Swamps, where they maintain themselves the whole winter long, and are not fetch'd home till the spring...and many of the poor Creatures perish in the Mire....These people live so much upon Swine's flesh, that it...makes them extremely hoggish in their Temper, & many...seem to Grunt rather than to speak."

Many Southerners didn't want to be Anglicans. This is a Lutheran church.

North Carolina was different. It had tough land to tame, so its settlers were apt to be free-spirited small farmers. Many were outcasts and religious dissenters from aristocratic Virginia. In Virginia the rich landholders were in control, and they were Anglicans. People who wanted to join the new religious sects—the Baptists and the Methodists—were often persecuted in Virginia. Some moved south to North Carolina.

North Carolina may have been the most democratic of all the colonies. Generally, North Carolinians minded their own business and left their neighbors alone, which may be why pirates made the North Carolina coast a base for their adventuring. Although some said it was because the pirates paid the North Carolina governor to leave them alone. That could be true, because it was finally a force from Virginia that got Blackbeard—the most famous and ferocious of the pirates. He braided his great black beard into pigtails, wove ribbons into them, and then hung smoking pieces of rope from his hat.

Right: how an artist imagined Blackbeard and his ruffians enjoying themselves at their Carolina hideout. There were women who liked piratical life too.

At night the pigtails looked like coiling snakes, and the burning rope gave his face a glowing, eerie look. It was enough to scare anyone, and it did scare a lot of sailors.

Anne Bonney scared them too. She was a tough pirate who sailed the Caribbean and Carolina coast, and had no trouble terrorizing seamen.

In 1677 some North Carolinians rebelled against England. They didn't like England's Navigation Acts, which forced them to pay taxes to England on goods sold to other colonies. In other words, if a North Carolina tobacco grower sold some of his tobacco to a merchant in Boston, he was supposed to pay a tax to England. Did that make any sense? The colonists didn't think so.

Genuine women pirates: Mary Read (left) and Anne Bonney. They were friends who survived a death sentence.

Some North Carolinians refused to pay. They even set up their own government and tried to get free of England. Sorry, said the British. They arrested the leader of the rebellion—a man named John Culpeper—and called him a traitor, but spared his life.

One hundred years later, in 1776, people in all the colonies were angry about English taxes. The colonists would unite, as Popé had united the Pueblo Indians—and...well, you'll see what happened.

The Death of Blackbeard

In 1718 Blackbeard stationed his ship in Pamlico Sound off the North Carolina coast. No ship was permitted to pass through without paying the pirate a toll. Alexander Spotswood of Virginia sent an expedition south. Lt. Robert Maynard, encouraged by the offer of £100 reward, fought the infamous pirate in a fierce hand-to-hand battle. "Blackbeard received a shot in his body...yet he stood his ground and fought with great fury till he received five and twenty wounds....At length, as he was cocking another pistol...he fell down dead." Maynard, to prove he had killed Blackbeard, cut off the pirate's head, stuck it on his bowsprit, and sailed home.

37 Royal Colonies and a No-Blood Revolution

Sir Edmund Andros wouldn't even let the people of Cape Cod keep the oil they got from stranded whales. Whales were "royal fish."

I hope you don't mind, but here is some more English history. It is the year 1686 (what century is that?), and James II is king of England. You met him before, when he was the Duke of York and his brother, King Charles II, gave him New York and New Jersey.

Now Charles is dead and James wears the crown. Unfortunately, James is not the nicest of kings. And it doesn't help that he is a Catholic, because most people in England belong to the Church of England.

King James II would like to rule without Parliament. He wants to be an "absolute monarch." He believes, like his father Charles I and his grandfather James I, that kings have a "divine right" to rule. That means he thinks God wants him to be king. He would like to make England Roman Catholic again. He will learn that you can't force religion on people.

James tries to take charge in America, too. Before his time, the English kings hadn't bothered much with America. The colonists were mostly left alone. Now, King James wants to change that, so he sends Sir Edmund Andros to New England. Andros is a tough, take-control person who tries to tell the Puritans how to run their colony.

It is hard to tell a Puritan anything. The New England Puritans came to America to run their own affairs, and they don't want an outsider making rules for them—especially Andros. They think he is a bully.

Finally, a mob goes after Andros. He is scared. You would be, too—mobs are dangerous. Andros puts on a woman's dress and tries to escape—oops!—he forgot to change his boots. Women don't wear heavy boots. Andros is spotted and captured. The Bostonians cool down and ship him back to England.

Do you think that the English Bill of Rights served as an inspiration for what was to come in America?

The king is annoyed with the troublesome colonists in America. They won't behave as they are told. But now King James has too many problems to worry much about the colonies. In England things are heating up. The English people have had enough of James II. They have another revolution—a civilized revolution. "You're not king anymore, James," is what they say. Since James isn't killed—as his father was in the English Civil War of 1649—this revolution, in 1688, is called the Bloodless or Glorious Revolution. It proves that revolutions don't have to be violent.

James's Protestant daughter, Mary, and her Dutch husband, William of Orange, are asked to be rulers. They are the couple the Virginians honored when they named the College of William and Mary.

The revolution is glorious because the English people make a deal with the new King and Queen. They insist that Parliament have more power than the monarchs. They also demand a Bill of Rights for the people. It is a terrible time for absolute monarchs, but a great moment for freedom. In America, the colonists find it inspiring.

Read that last paragraph again. It's important to remember that *the glorious revolution gave parliament more power than the king. HOORAY!*

Do you think that makes things better for the colonies? Well, sometimes parliaments can be as pesky as kings.

When the English parliament begins to make rules for the American colonies, the colonists will get annoyed. The Americans will complain that Parliament is too powerful, especially when Parliament ignores Massachusetts's beloved charter and makes the state a royal colony with a royal governor.

The grumbling that begins in New England in 1691 will start the colonists thinking about having their own glorious revolution. They will have one—but it will not be bloodless.

A king and queen who helped to make a peaceful revolution.

In 1687 Connecticut's charter was hidden in this Hartford oak tree to keep it out of Andros's hands. The tree blew down in 1856 and everyone who could cut a piece for a souvenir.

38 A Nasty Triangle

Men are loaded onto a slaver (slave ship) and chained together at the ankles.

You may be getting the idea that the United States began as a collection of settlements that were not much alike. And you are right.

South Carolina wasn't like Pennsylvania, and Maryland wasn't like Connecticut. The people who founded the colonies had a lot to do with those differences, and so did the conditions of the land.

Massachusetts had a special problem because of its rocky soil and cold climate. It was tough being a farmer in New England, but New Englanders were tough people who liked challenges. So they did farm, although for many it was "subsistence farming." That means they grew enough for themselves; they didn't usually have extra crops to sell. A few New England farmers were able to sell their farm products abroad but, mostly, New England's land just wasn't right for large farms—or plantations—like those in the South.

And when it came to industry, the British made things difficult. They wouldn't let the colonists manufacture goods that competed with English goods. You can understand why that caused some grumbling.

New Englanders had to find ways to earn a living. Fishing was one way. Cod became New England's gold, just as tobacco was Virginia's.

The people in this shed are cleaning and drying codfish, which was (and still is) an important trading commodity for New England.

The New England lumber mill cuts and splits the logs and then spits them out straight into the river so they can float down to the towns and shipyards that need them for building houses and boats.

The Puritan settlers caught codfish and then salted and shipped and sold the fish in Europe or the Caribbean Islands. In order to do that, they needed ships. So they became shipbuilders. To make ships they needed lumber. So they harvested timber and began selling wood and wood products. They became merchants carrying goods around the world. Yankee ships were familiar sights in Singapore and Rangoon and Bristol. And New England boys, who hung around the wharfs, got a chance to touch Dutch coins, Chinese silks, or fruit from Spain. They heard tales of adventures in Tripoli and Jamaica and dreamed of becoming skippers and going to faraway places themselves.

Soon Yankees were trading all kinds of things. They might take their salted cod to Barbados and trade it for cane sugar. Then they'd go to Virginia to pick up tobacco. They'd take the tobacco and sugar to England and trade them for cash, guns, and English cloth. Then on to Africa where they exchanged the guns and cloth for men, women, and children. From there it was back across the Atlantic Ocean to the West Indies where the people were sold into slavery. Finally they sailed home to New England (or, sometimes, New York or Annapolis). All that was called the triangular trade. It made some people very rich.

Picture a triangle—a long one. Do you have three points in your mind? Now stretch the triangle across the Atlantic Ocean. Put one

Between 1526 and 1870, nearly 10 million slaves were shipped from Africa to: Europe (175,000); Spanish America (1,552,000); Brazil (3,647,000); British Caribbean (1,665,000); British North America and United States (399,000); French America (1,600,000); Dutch America (500,000); Danish West Indies (28,000).

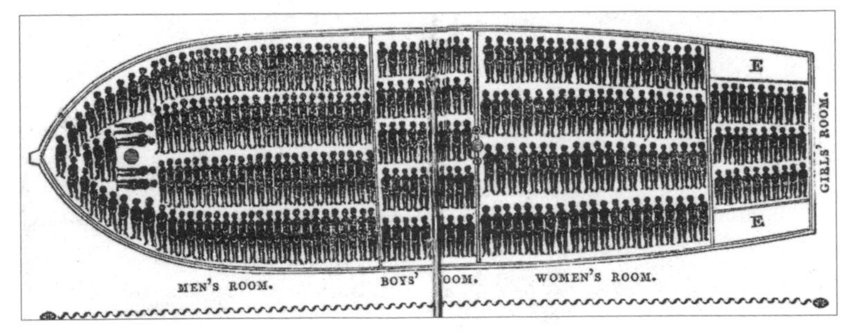

A famous diagram of a slave ship that shows how the people were stowed, each in a space maybe 15 inches by six feet.

point on the New England coast, another in Africa, and the third in the South Carolina. Are you sure you have that clearly in your mind? Now imagine a boat sailing along that triangle, from New England to Africa to South Carolina, and back to New England.

Stretch another triangle across the Atlantic. This one can start in England, go to Africa, and have a third point in Virginia. The Atlantic Ocean was once filled with ships sailing triangular routes. Most of them included a stop in the West Indies. (They were very jagged triangles.)

Let's pretend a triangle is starting at Newport, Rhode Island—where many did. You can watch as a ship is loaded with rum and guns. (Rum is an alcoholic drink made from sugarcane.) The ship heads for Africa, where the rum and guns will be traded for African people.

The Africans have been captured by enemy tribesmen and sold to African slave traders. The slave traders bargain with the New England boat captain, who buys as many people as he can squeeze on his ship. Some of the captives are children, kidnapped from their parents.

Olaudah Equiano was one of those children. He was 11 in 1756, when he was captured in Benin. He was the youngest of seven children, a happy boy in a loving home. Like many other prosperous African families, his family had slaves. Imagine that you are Olaudah as you read his words:

One day, when all our people were gone out to their works as usual, and only I and my sister were left to mind the house, two men and a woman got over our walls, and in a moment seized us both; and without giving us time to cry out or to make any resistance, they stopped our mouths and ran off with us into the nearest wood. Here they tied our hands, and continued to carry us as far as they could, till night came on, when we reached a small house, where the robbers halted for refreshment and spent the night.

Olaudah and his sister are taken on a long journey, separated, and sold. He is

Born Free

In a graveyard at Concord, Massachusetts, stands a stone carved thus:

God wills us free—
man wills us slaves
I will as God wills:
Gods will be done.
Here lies the body of
John Jack
A native of Africa
who died March 1773,
aged about sixty years.
Tho born in the land
of slavery
He was born free:
Tho he lived in a land
of liberty
He lived a slave
Till by his honest
tho stolen labours
He acquired
the source of slavery
Which gave him his freedom:
Tho not long before
Death the great Tyrant
Gave him his final
emancipation
And put him on a footing
with kings.
Tho a slave to vice
He practised those virtues
Without which kings are
but slaves.

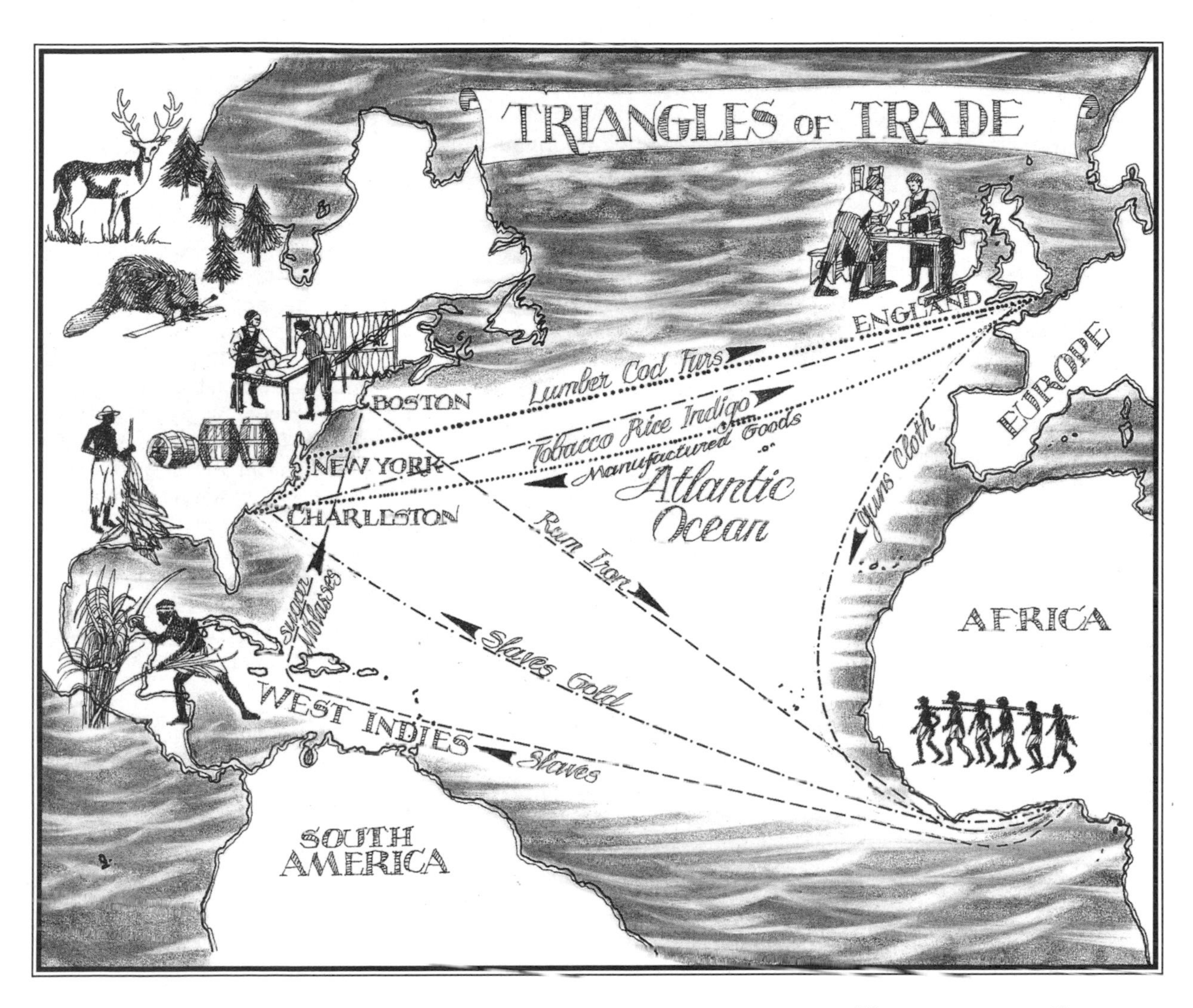

passed from person to person, staying a month here, a few weeks there. Olaudah sees many parts of Africa and has many adventures. He tries to run away but is unsuccessful. Then, for the first time in his life, he sees the ocean.

> *The first object which saluted my eyes when I arrived on the coast was the sea, and a slaveship, which was then riding at anchor and waiting for its cargo. These filled me with astonishment, which was soon converted into terror.... When I was carried on board I was immediately handled, and tossed up to see if I was sound, by some of the crew; and I was now persuaded that I had got into a world of bad spirits, and that they were going to kill me.*

Olaudah is tossed below deck, where the smell is so bad he be-

There was more than one trade triangle. But all of them were tied up with slavery, and slavery was tied up with them. Most of the people who made money out of slavery didn't want to see it come to an end.

comes sick and cannot eat. When he refuses food he is tied down and beaten. Frightened, he is at first unable to talk to anyone because the other Africans do not speak his language. Finally he meets some chained men who speak the language of Benin.

> *I asked them if these people had no country, but lived in this hollow place [the ship}. They told me they did not, but came from a distant one.... I then asked where were their women? Had they any like themselves? I was told they had. "And why," said I, "do we not see them?" They answered because they were left behind. I asked how the vessel could go? They told me they could not tell; but that there was cloth put upon the masts by the help of ropes I saw, and then the vessel went on; and the white men had some spell or magic they put in the water, when they liked, in order to stop the vessel. I was exceedingly amazed at this account, and really thought they were spirits.*

Olaudah learns that he is being taken to the white men's country to work.

> *I then was a little revived, and thought, if it were no worse than working, my situation was not so desperate: but still I feared I should be put to death, the white people looked and acted, as I thought, in so savage a manner; for I had never seen among any people such instances of brutal cruelty; and this is not only shown towards us blacks, but also to some of the whites themselves. One white man in particular I saw, when we were permitted to be on deck, flogged so unmercifully with a large rope near the foremast, that he died in consequence of it; and they tossed him over the side as they would have done a brute. This made me fear these people the more; and I expected nothing less than to be treated in the same manner.*

Olaudah describes the scene below deck, where people are packed so closely they can hardly turn over. The smells, he says, are "loathsome." Women shriek, the dying groan, all is "a scene of horror."

> *One day, when we had a smooth sea and moderate wind, two of my wearied countrymen, who were chained together (I was near them at the time), preferring death to such a life of misery, somehow made through the nettings, and jumped into the sea.*

Olaudah is taken to Barbados in the West Indies where he is sold. His story is different from most. He will go to sea as a slave, have many adventures, learn to read, and write his autobiography.

Africa: The Unknown Continent

North Africa was the only part of Africa the Europeans knew. And, of North Africa, they knew only the lands that touched the Mediterranean Sea. Those lands were rich in history. Egypt had once seen a civilization that produced pharaohs, pyramids, and a sphinx. That was more than 2,000 years before the birth of Christ. Much later, in the 9th century C.E., another great civilization flourished in North Africa. It was Islamic (also called Muslim), and it began in Arabia and spread to Morocco and Spain and Mediterranean Europe.

But what of the rest of the huge African continent? What was it like? Europeans knew almost nothing of it—although there were wild rumors of rich kingdoms and of seven lost cities.

There was a barrier of sand that kept the Europeans from learning much: the vast Sahara desert. A few people were able to cross those blazing desert sands. They were African or Arab traders who traveled from oasis to oasis carrying gold and slaves from lands to the south. It was a very dangerous journey.

Why didn't the Europeans just sail down the African coast and discover for themselves where that gold was coming from? They wanted to, but until the 15th century they couldn't do it. You see, their boats were powered by men with oars. Those boats were fine in the calm Mediterranean, but

they weren't safe in the rough Atlantic waters. It was not until the 15th century that the Europeans developed sailing technology—learned from the Muslims—that allowed them to build caravels that could sail into the wind.

Then they got up their courage and began going down the coast, farther and farther, until the Portuguese sailor Vasco da Gama rounded the tip of Africa in 1498 and went on to India. Now, perhaps, the Europeans could stop along the African coast and visit the rich grasslands and forests below the Sahara. They could—if the Africans would let them.

But the people who lived in Africa weren't anxious to have outsiders come and explore or settle. They welcomed the Europeans as traders with goods they could use, but that was all. The Europeans couldn't land, take over, and mine the continent's gold, as they did in America. The African warriors were too strong. The Africans let the Europeans build a few trading posts, but nothing more. Besides, there were African diseases that frightened the Europeans.

So the Europeans didn't learn much about the peoples and cultures of the African continent. For them, it remained mysterious. If they had been able to explore, they would have discovered as much variety in Africa as in Europe. There were sophisticated empires and primitive cultures. There were sculptures cast in bronze and gold and useful iron objects hammered by village blacksmiths and cotton that weavers turned into handsome fabrics.

On the west coast of Africa, near the continent's bulge, the three great kingdoms of Ghana, Mali, and Songhai rose and fell between the time of the Roman Empire and the settling of North America. The ruler of Mali, Mansa Musa, was a Muslim who in 1324 took so much gold to Mecca (the center of the Muslim world) that the world price of gold tumbled. In the 15th century, Timbuktu, the leading city of Songhai, was renowned for its schools and wise men. A visitor wrote of the city's "great store of doctors, judges, priests, and other learned men that are bountifully maintained at the King's cost...and hither are brought diverse manuscripts or written books...which are sold for more money than any other merchandise."

These bronze sculptures were carved in Benin, the country in West Africa that Olaudah Equiano was kidnapped from.

But by the time the sailing ships were able to call at West African ports, it was trade alone that interested both peoples. The Europeans had guns, iron, cloth, kettles, and mirrors that were wanted in Africa. The Africans had workers—healthy, hardy people—who were wanted to grow crops and mine and settle the place the Europeans called a New World.

And so men and women would be traded into slavery by people, on both sides of the Atlantic, who didn't seem to worry about the consequences of their actions.

He will take a European name. It is Gustavus Vassa.

Many Africans are sent to Virginia, where they are traded for tobacco. Some are exchanged for sugar and molasses in the West Indies. Others are traded for rice in South Carolina. Then the ships head back to their home ports.

In Newport, Rhode Island, where we started this voyage, the sugar and molasses are turned into rum—and the triangle begins again. That is the way the terrible triangular trade works. Every colony is a part of it. English ships carry the greatest numbers of Africans into slavery.

In the colonies, laws are soon passed that attempt to take away the blacks' humanity. The Virginia Black Code says that slaves are property—not people. New York law says runaway slaves caught 40 miles north of Albany—on the way to Canada and safety—are to be killed.

Remember when the first black people arrived as indentured servants at Jamestown? In 1725 about 75,000 blacks are living in the American colonies. By 1790 there are more than 10 times that number.

A tobacco trader from New Amsterdam.

39 Four and Nine Make Thirteen

This man has been brought to debtors' prison by his creditors (the people he owes money to), to stay there until he finds a way to pay them.

Debtors' prisons in England stuck around for more than a century after Oglethorpe. The author Charles Dickens used his writing skills to help get rid of them. When Charles was a boy, his father had been thrown in debtors' prison—so he knew just how dreadful it was. He described those prisons in some of his novels—such as *Little Dorrit* and *Our Mutual Friend.* Those are long books, but they tell wonderful stories.

Tomochichi, leader of a Creek town near Savannah, welcomed Oglethorpe.

If you've been counting, you know I've talked about 12 colonies. Can you name them? Cover the next sentence and see if you can.

Here they are: Massachusetts, New Hampshire, Connecticut, Rhode Island, New York, New Jersey, Pennsylvania, Delaware, Maryland, Virginia, North Carolina, and South Carolina.

Finally, like a tail at the end of a kite, along came the 13th colony. Do you know what it was named? Here's a clue: It was founded in 1732, when KING GEORGE II was on the throne of England.

That wasn't hard: it was Georgia.

Georgia's beginning was noble, not because of birth or wealth, but because of a noble idea. Unfortunately the idea didn't work out. Still, it was inspiring.

James Oglethorpe, who planned Georgia, wanted to solve a terrible problem. People in England who couldn't pay their bills were thrown into debtors' prisons. Once they were in those jails—and they were awful places—they couldn't work or earn money, and so they had no way to pay their debts. If they were lucky, a relative or friend came up with the cash. Otherwise they just stayed there. Many died in prison.

Oglethorpe decided to found a colony where debtors could go instead of going to jail. He wanted to make it a place where people could lead ideal lives. So he had laws passed for Georgia that made drinking liquor and keeping slaves illegal. He wanted Georgians to live on small farms, not big plantations, and he wanted them to do their own farming. He brought experts from Europe to teach them how.

In 1734 settlers began building the town of Savannah. The tent under the tall trees in front of the houses was James Oglethorpe's.

These Lutherans left Austria to find freedom to worship in Georgia. The verse says their only companion is the Gospel—having left their homeland, they are in God's hands.

Oglethorpe helped to plan a handsome capital, Savannah, a city with beautiful parks and fine public squares. It was a shame his idea didn't work out. Not many debtors wanted to come to the wilds of Georgia. To some, even prison seemed safer.

Those who came were much like the settlers in the other colonies: a mixture of peoples and religions. Anglican men and women came, German Lutherans came, Catholics came, Jews came, and so did Scotch Presbyterians.

They soon discovered that Georgia was full of Indian villages. When the first settlers arrived, they found thousands of Indian mounds. The mounds were sacred sites from the Native American past. (If you get into a helicopter and fly near Eatonton, Georgia, you can see one of those earth mounds. It is shaped like an eagle with wings that spread out for 120 feet, about the size of four average classrooms.)

The local Indians had bad memories of white people. Back in 1540, the Spanish conquistador, Hernando de Soto, had marched through Georgia with his army. De Soto was so cruel and evil that the Indians were still telling stories about him 200 years later.

But Oglethorpe was a fine and honorable man, and the Indians learned to trust him. They made many peace treaties with him and kept all of them.

It was the settlers who gave Oglethorpe problems. They wanted to drink liquor and have slaves, and eventually they won out. When Oglethorpe tried to force his laws and "good ideas" on others, it just didn't work.

Besides, life in Georgia wasn't easy. Spaniards and pirates gave the settlers a hard time. Pirates roamed along the Georgia coast, capturing ships of all nations. Spain controlled Florida and said Georgia and the Carolinas were also her territories. Because of that, there were constant border fights. When the Spaniards attacked, the Georgians were able to fight them off—luckily for them, because none of the other colonies helped out. It was a while before the colonies thought of uniting and helping each other.

James Edward Oglethorpe, founder of Georgia.

Oglethorpe lost all his money trying to establish Georgia, but he finally gave up. Georgians wanted to have rice plantations and slaves and the king's government, and that is what they got. The king made Georgia into another royal colony (in 1752). That made eight royal colonies: Virginia, Massachusetts, New Hampshire, North Carolina, New York, South Carolina, New Jersey, and Georgia. Each had a royal governor, appointed by the king.

Rhode Island and Connecticut had charters that allowed them to govern themselves. Their assemblies (congresses) picked their governors. Maryland, Pennsylvania, and Delaware were proprietary colonies. They were owned by individuals—the Calverts and the Penns.

All the colonies had assemblies of local leaders who made most of their laws. Later England would regret the freedom she gave the colonies and try to take some of it back. That would lead to big trouble.

But in the mid-18th century most Americans were happy to be part of the mighty British empire. The 13 colonies were like 13 children of a kindly and faraway parent. Each colony seemed to be a tiny nation, with its own government, its own habits, and its own religious ways.

They were different—one from another—but in some ways they were all a bit like Europe. All had a measure of European class society. So, if you really wanted to be independent, if you wanted to be the equal of anyone, the place to be was on the frontier. There, it was your intelligence and your strength that made you a leader. On the frontier no one cared if you were Puritan or Anglican. It didn't make a difference if your father was a lord or a pauper. Could you be depended upon? Did you tell the truth? Could you shoot straight? Were you brave? That's what mattered on the frontier.

The frontier offered something even more important: *land*. Land meant everything in a society that lived by farming. *Owning land made you feel really free.*

People were arriving in America every day. Much of the land on the East Coast was taken. The frontier was where these newest Americans were going. They were heading to the tree-thick mountains that bordered the coastal plains—and then on, over those mountains.

A famous English poet, Alexander Pope (he wrote that "A little learning is a dangerous thing"), composed a short verse, or couplet, about Oglethorpe (they were friends):

One driven by strong benevolence of soul,
Shall fly like Oglethorpe from pole to pole.

What does that mean?

James Oglethorpe (in black) went back to London to introduce some of the Indians who had sold him land and befriended him to the trustees of his new colony.

40 Over the Mountains

Look at the map of the 13 colonies on page 90 to see how the colonies hug the East Coast. You will also see the Appalachian mountains running all the way down the left boundary of the colonies, blocking the way west.

A family traveling west lights a fire. Imagine cooking dinner at night in the rain and the unknown.

Look at a map. Notice: all the early English-speaking settlements are on or near the East Coast. That's not surprising, since the settlers came by boat from Europe. Most of them stayed near the place they landed.

Yet, almost from the first, a few settlers wanted to know what the land was like to the west. They came back with tales that made others want to go, too. Some went because they thought they would find the Pacific Ocean; some went west to trade with the Indians; a few were criminals who were chased away from the settlements. Most didn't get far.

Look at the map again, and you'll see why. Do you see that long strip of mountains that runs from north to south and makes a spine down the back of the eastern United States? Those are the Appalachian Mountains, and they stretch from Quebec to Alabama.

The Appalachians are the oldest mountains in our country. They are tall, but not as tall as some of the far western mountains. The Appalachians have been rounded and smoothed by millions of years of wind and rain.

Don't be confused if you hear of White Mountains in New Hampshire, Green Moun-

The Swedish introduced loghouses to the New World. They were just right for places that had a lot of trees and not much else.

Can you guess the remote location of this lonely log cabin, surrounded by thick forests? It's a few miles from Baltimore, Maryland. In the 18th century you didn't have to go far to be out in a wilderness.

tains in Vermont, Catskill Mountains in New York, Allegheny Mountains in Pennsylvania, and Blue Ridge, Clinch, Shenandoah, and Great Smoky Mountains in the South. All these mountains are part of the Appalachian range.

Water from Appalachian streams and snows flows in two directions. Some water goes into rivers that drain into the Atlantic; some goes to rivers that drain into the Gulf of Mexico. The spot high on the mountains where the waters change direction is called a divide.

Have you ever climbed a mountain? It isn't easy—even today, when there are clear trails. Imagine what it was like for the first trailblazers. The pioneers who went into Vermont couldn't even take horses. They had to go on foot, because the mountains were so thick with trees that no grass could grow. So there was nothing for horses to eat. Those explorers had to go with their axes, cut down trees, plant seeds, and wait for a crop before they could bring animals.

The trailblazers were first, but they were soon followed by people looking for land to farm. Usually mountain land isn't easy to farm, so they didn't stay in the mountains. They went on, over the Appalachians, to the valleys or the flat lands beyond. There was no government where they went, so they were on their own. They had to fight Indians and other settlers for their land. They were a tough breed, those early over-the-mountain people. They had to be in order to survive.

Oh, Shenandoah

Alexander Spotswood, the Virginia governor we met earlier, was a big booster of westward expansion. In 1716 he set out with an expedition into Virginia's mountainous wilderness. They traveled 200 miles, named the Blue Ridge Mountains (a trick of the light makes them look blue), and reached the Shenandoah valley. When they came home, Spotswood gave each member of the expedition a tiny gold horseshoe (because their horses had to be shod to manage the rocky mountain paths). Afterward, they were known as the "Knights of the Golden Horseshoe."

Most didn't have time to write of their adventures, so we have to listen to others to try and understand what it must have been like. A famous French diplomat named Talleyrand visited the United States at the end of the 18th century. He wrote of a trip west. Talleyrand had a guide, and he went on horseback. This is what he saw:

> *I was struck with astonishment; less than 154 miles away from the capital [Philadelphia] all trace of men's presence disappeared. Nature, in all her primeval vigor, confronted us. Forests as old as the world itself...here and there the traces of former tornadoes that had carried everything before them. Enormous trees, all mowed down in the same direction, extending for a considerable distance, bear witness to the wonderful force of these terrible phenomena....To be riding through a large wild forest, to lose one's way in it in the middle of the night, and to call to one's companion in order to ascertain that you are not missing each other; all this gives impressions impossible to define....When I cried, "...are you here?" and my companion replied, "Unfortunately I am, my lord," I could not help laughing at our position.*

Talleyrand, who was one of the French emperor Napoleon's most important statesmen, is also supposed to have said: "The United States has 32 religions but only one dish." He didn't think much of Yankee cooking.

Do you use the expression "I'm stumped" when you're stuck about what to do or think? In the days when westbound wagons had to travel terrible roads like this one—where tree stumps stuck out of the ground—if you hit a stump, you were in trouble—"stumped."

41 Westward Ho

Daniel Boone was a "long hunter," so called because he stayed in the wilderness for weeks.

The British tried to stop them. They didn't want settlers moving west. The British were in charge, and that meant they had to keep order. If the settlers moved into Indian territory, someone had to worry about protecting them from Indian attack. Someone had to protect the Indians from rifle-happy settlers. Someone had to make treaties with the Indians and do some governing.

That cost money. Britain would have to build forts in the western territories. She would have to send soldiers to man those forts. She would have to send governors. Parliament told the colonists to stay out of Indian territory.

But there was no stopping them. Some people want to go where no one has gone before. Some people are born explorers. Some people are just restless. The Americans who headed for the frontiers were all of those things—and more.

They were people who were looking for a good life, and they hadn't found it on the East Coast. By the middle of the 18th century the best land in the East was taken, and society was already in place. These people weren't content to sit around and be second-best. They were the kind who would sail across a dangerous ocean to try their luck in a new world. That kind of person was willing to risk everything for adventure or the dream of a better life.

So they headed out—and you had to be brave to do that—into the unknown wilds. They were called frontiersmen and frontierswomen, trailblazers and pathfinders. They lived in the woods, shot the food they ate, and made their own clothes. They learned from the Indians and from each other. One of them was a man named Daniel Boone.

Daniel Boone was born in 1734, two years after James Oglethorpe

Do you know why "buck" is another word for *dollar*? It's because in the days of Daniel Boone and the Long Hunters, a buckskin was worth about a dollar—so the money was named for the deer that it bought.

Daniel was not the first Boone to seek adventure. His ancestors were Normans—soldiers and farmers from French Normandy—who conquered England in the 11th century and settled there because of the good land and opportunity. To them, then, England seemed a frontier on the edge of the world.

Women and children from settler families were sometimes kidnapped by Indians during the border battles over territory. Daniel Boone's own daughter, Jessica, was carried off in 1776, but he managed to rescue her. This is an artist's rather romantic idea of the event.

founded Georgia. Oglethorpe would have liked Daniel Boone, and John Smith would have like him, too. Daniel Boone didn't look for others to do his work. He grew up in Pennsylvania, with Indians for friends, and he learned their ways along with the ways of the European settlers. That was an interesting combination, and it was creating a new people, a people who were wholly American.

Daniel Boone's grandfather, George Boone, had lived in England, where he was a weaver and an independent thinker. He was both cautious and daring, a man who would do what he intended and do it well. When he was 36 years old George Boone left the Church of England and joined an outcast religion that some people called Quaker. Boone knew it as the Society of Friends.

Being a Quaker in England was hard, as we've already seen. Quakers were persecuted, but George Boone stood up for his beliefs.

Then he heard about William Penn's colony in America, where Quakers and everyone had equal rights. He also heard stories of abundant land cheap enough for anyone to buy, and that seemed almost too good to be true.

As I said, Boone was cautious. He had a wife and nine children. Before he moved his whole family, he sent his three oldest children to the New World to look around. This was what George Boone learned in a letter from Pennsylvania:

> *Because one may hold as much property as one wishes and pay when one wishes, everybody hurries to take up property. The farther the Germans and English cultivate the country the farther the Indians retreat. They are agreeable and peaceable....in summer one can shoot a deer, dress the skin, and wear pants from it in twenty-four hours.*

It sounded good to him. In 1717, when he was 51 years old—an old man in those days—George Boone left England and sailed for Pennsylvania. Boone's grandson, Daniel, became the most famous of all the American frontiersmen.

Daniel Boone's first adventures came in 1755, during the French and Indian War (which you'll read more about in Book 3). He fought on the side of the British, as did George Washington. Washington was an officer; Boone was a wagon driver. Sitting with soldiers around the campfire, young Boone heard John Finley, a fur trader, tell stories of a land

Have you ever seen a picture of Daniel Boone in a coonskin cap? Well, those pictures were drawn by people who didn't know Boone. He wore a broad-brimmed hat to protect his eyes from the sun.

the Indians called *kentake*, which means "meadowland." Finley said it was the most beautiful land he had ever seen, filled with high grasses, birds, buffalo, deer, and beaver. It was an Indian hunting ground, and the Indians didn't want the white men to find it.

Daniel Boone got excited about those meadowlands over the mountains. As soon as the war was over, off he went, with a sack of salt around his neck, an ax in his belt, and a rifle over his shoulder. But he never found an easy way to get over the mountains.

Then, one day, a peddler knocked on his door. It turned out to be John Finley. Finley told him of an Indian trail through the mountains.

Boone searched until he found that trail. It led to a hole through the mountains, called a gap, and that led to the rich grasslands of Kentucky. In 1775 Daniel Boone and 30 woodsmen turned that Indian trail into a road that families could travel with wagons and animals. It was called the *Wilderness Road,* and it went for 300 miles. By 1790, almost 200,000 people had gone west on the Wilderness Road.

Three years after Daniel Boone first reached Kentucky, settlers were already pouring in. A man named James Harrod was laying out a town named after him. Pretty soon there was a town named after Boone, too—Boonesboro.

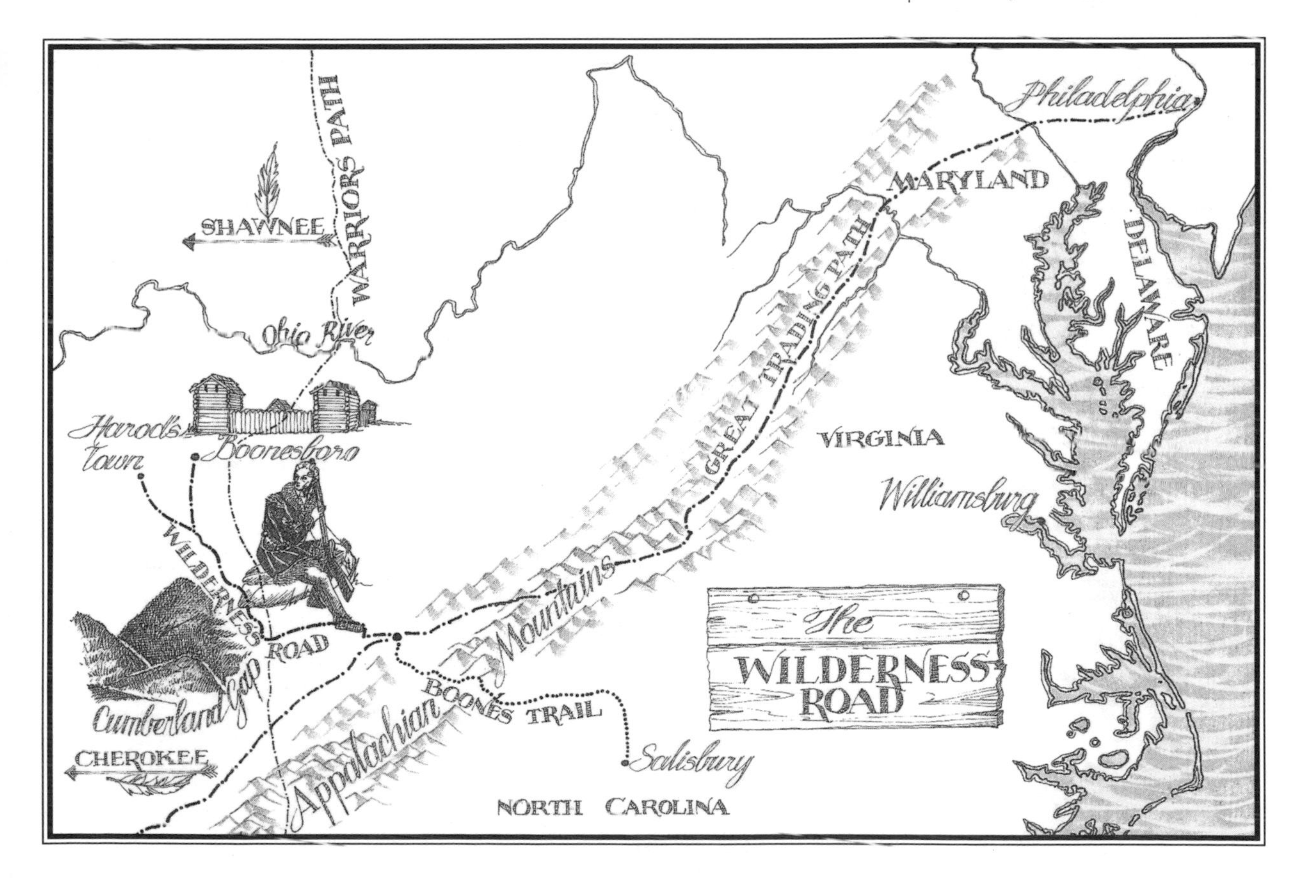

This bloodthirsty picture of *An Attack of the Indians upon Dan'l and Squire Boone & John Stewart* helped spread the legends told about Boone after his death. (Squire Boone was Daniel's father.)

But Daniel Boone just couldn't stay settled. He was always off on one adventure or another. The Indians didn't want Boone or any whites in Kentucky. Then they hoped the settlers would at least all stay in one place. They soon learned that there was no way to keep Daniel Boone from wandering. He was a natural-born explorer, and, like a cat, he had nine (or maybe more) lives. His wife was told he was dead—more than once—and a few times he was close to it, but he always survived, whether his attackers were animals or people.

Sometimes he ran backward through the woods, and Indians following his trail went the wrong way. Once he swung 40 feet on a grapevine so Indians would lose his trail. The stories about Boone's adventures grew and grew and grew, and some of them were true: stories about how he was captured by Indians, about how he rescued his daughter from Indians, about a buffalo stampede, about how he was adopted into a tribe. Each time the stories got told, they were bigger and better.

It was said that Daniel Boone could shoot a flea off the nose of a bear. Do you think that could be true?

We would know for sure if it weren't for a canoe accident. When Daniel was an old man, he told his grandson about all his adventures. His grandson wrote them all down in a book. Everyone who knew Daniel Boone agreed that he was honest, so that book must have told the real story about him. But the book was in a canoe when the Boone family moved; the canoe tipped; and the book was lost.

We do know for sure that when Daniel was a boy one of his best friends was named Abraham Lincoln. Many years later, that Abraham's grandson became a famous president. President Lincoln was proud that his grandfather had known Daniel Boone, because Daniel Boone was a real American hero.

42 The End—and the Beginning

This painting of George Washington was made when he was a young colonel in the Virginia militia—at the very beginning of what was to be an extraordinary career.

Congratulations! You have finished Book 2 of *A History of US*. The 13 colonies have been formed, and are doing very well!

We have traveled into the 18th century, and those early days of hunger and hardship are now only a memory. Life is beginning to be comfortable for the settlers in this New World. It still takes courage to cross the dangerous Atlantic and begin again on a continent that is reshaping itself. But, just as the birth of a baby offers promise and hope, so, too, this vast, inviting land offers hope and possibility to all who are tired of the Old World's ways.

Many people have come to America because that Old World held out no opportunity for them. Most think they have found paradise in this land of sweet-smelling flowers, abundant game, and fertile soil. In England, men are shot for hunting the king's deer. In America, deer, beaver, birds, and fish are for everyone.

Some of those who are here have a powerful dream of a free country where everyone is treated fairly. The dreamers will have the opportunity to test their ideas. They will turn 13 colonies into a nation "conceived in liberty, and dedicated to the proposition that all men are created equal."

If you keep reading our history, you will learn about the making of that nation—our nation. You will meet people like George Washington and Thomas Jefferson. You will learn more about Benjamin Franklin and Daniel Boone. The story will include war, adventure, heroism, villainy, and love. This is the end of Book 2. But don't stop now. There is much excitement ahead!

Chronology of Events

WARNING: CHRONOLOGIES CAN BE DANGEROUS!

Medical reports tell of a disease—called *date-itis*—with these alarming symptoms: rapid heartbeat, heavy sweating, and frequent moaning. This disease is caused by an allergy to dates. Teachers need to be aware that this has nothing to do with dried fruit! This not-very-rare disease affects people who hate to memorize dates. If the disease is treated skillfully, recovery is possible.

NOTICE TO POSSIBLE VICTIMS OF THE DISEASE:

Before reading this page: place a glass of water in your right hand, an aspirin in your left, and read rapidly. Chances are good that you will survive, and you might learn something, too.

1607: three ships sent by the London Company land at Cape Henry, Virginia. Captain John Smith and others found Jamestown

1609: Henry Hudson sails up the Hudson River

1609–10: all but 60 of the 500 settlers in Jamestown die during the Starving Time

1610: Santa Fe founded in New Mexico

1612: settlers plant tobacco in Virginia for the first time

1614: Pocahontas, daughter of Powhatan, marries John Rolfe in Jamestown

1619: the first African slaves arrive in Virginia, and the first representative government of the European colonies begins with Virginia's House of Burgesses

1620: The Pilgrims sail from England to Cape Cod on the *Mayflower*. They make a plan of government called the Mayflower Compact

1621: Massasoit, leader of the Wampanoag Indians, establishes peace with the Pilgrims

1622: Opechancanough leads a great massacre of English settlers around Jamestown

1626: The Dutch buy Manhattan island from the Indians

1630–40: The Puritans keep arriving in New England and spreading through the Massachusetts Bay Colony and beyond

1632: Lord Calvert founds Maryland

1633: first school in the North American colonies founded in New Amsterdam

1636: Roger Williams founds Rhode Island; Thomas Hooker moves to Connecticut; Harvard College founded

1649: King Charles I of England is beheaded; Parliament rules

1660: Charles II restored to the English throne; Mary Dyer hanged in Massachusetts

1664: the British capture New Amsterdam and rename it New York

1665–66: London attacked first by the Great Plague and then by the Great Fire

1670: Charleston founded in South Carolina

1675–76: Massasoit's son Metacom fights New England's colonists in King Philip's War

1677: North Carolinians rebel against English taxes

1680: Pueblo Indians led by Popé attack the Spaniards and drive them out of Santa Fe

1681: William Penn founds Pennsylvania

1686: King James II sends Sir Edmund Andros to control New England colonists

1688: England's Glorious Revolution; James II deposed

1692: witchcraft trials held in Salem, Massachusetts

1699: Williamsburg becomes capital of Virginia

1706: birth of Benjamin Franklin in Boston

1732: James Oglethorpe founds Georgia

1734: birth of Daniel Boone

1775: the Wilderness Road open to pioneers

More Books to Read

History and Biography

Aliki, *The Story of William Penn*, Prentice-Hall, 1964. This book is very easy to read and has colorful pictures.

Franklin Folsom, *Red Power on the Rio Grande*, Follett, 1973. The story of Popé and the Native American revolution of 1680.

Jean Fritz, *The Double Life of Pocahontas*, Putnam, 1983. Biography as it should be written: exciting and informative.

Jean Fritz, *Who's That Stepping on Plymouth Rock?*, Coward McCann, 1975. Easy to read, fun, and good history, too.

Ann McGovern, *If You Lived in Colonial Times*, Scholastic, 1992. A clear, informative, colorful book. Part of a series that answers many questions about past times.

Marcia Sewall, *The Pilgrims of Plimoth*, Atheneum, 1986. Lovely paintings and a nice story.

Historical Fiction

Patricia Clapp, *Constance: A Story of Early Plymouth*, Beech Tree Books, 1991. This isn't exactly fiction, because it's the imagined diary of a real girl, Constance Hopkins, the daughter of one of the original *Mayflower* Strangers. It's written (very well) by one of her descendants, and it is as lively and fun as Constance herself.

Rachel Field, *Calico Bush*, Macmillan, 1931. Marguerite Ledoux is a 13-year-old French orphan, bound out for six years to a family who move to the wilds of Maine in 1743. This story, written over 60 years ago, is real, fresh, and exciting. It might even make you cry.

Paul Fleischman, *Saturnalia*, HarperCollins, 1990. Fourteen-year-old William is a printer's apprentice in 1681 Boston—but he's also a captured Narraganset Indian trying to find a link with his past. This story is very well written and researched—and quite thrilling.

Sally M. Keehn, *I Am Regina*, Dell, 1991. This is based on a true story about an 11-year-old girl who was kidnapped by Indians in 1755 and lived with them for nine years before being returned to her mother and their home in western Pennsylvania.

Ann Petry, *Tituba of Salem Village*, Harper & Row, 1964. Tituba, a slave from Barbados, is sold to a minister who moves to Salem, Massachusetts. This thoughtful book tells the story of the Salem witchcraft trials from Tituba's imagined point of view.

Elizabeth George Speare, *The Sign of the Beaver*, Houghton Mifflin, 1983. Matt is alone in the woods in Maine while his father fetches his mother and sister. He's befriended by an Indian boy, Attean—and both boys learn some surprising things about each other.

Elizabeth George Speare, *The Witch of Blackbird Pond*, Dell, 1958. Seventeenth-century Connecticut is the setting for this classic tale of suspense and romance. Highly recommended.

Elizabeth Yates, *Amos Fortune: Free Man*, Dutton, 1950. Born a prince in 1710 in Africa, At-mun is captured, sold in Massachusetts, and, as Amos Fortune, works in slavery until he buys his freedom at the age of 60. He dies a respected free man in New Hampshire, aged about 90. A splendid and true story.

Poems, Folktales, a Play, Songs, and Food!

Suzanne I. Barchers and Patricia C. Marden, *Cooking Up U.S. History: Recipes and Research to Share with Children*, Teacher Ideas Press, 1991. Recipes and information about American food. You can learn how to make hasty pudding, berry ink, hardtack, snickerdoodles, gumbo, and sopapillas.

Arthur Miller, *The Crucible*, Viking, 1953. This is a play about the Salem witchcraft trials by a famous playwright. It is not hard to read.

Scott R. Sanders, *Hear the Wind Blow: American Folk Songs Retold*, Bradbury, 1985. I love this kind of book. Stories and songs and explanations.

Alvin Schwartz, *Scary Stories to Tell in the Dark: Collected from American Folklore*, Lippincott/HarperCollins, 1981. Just what the title says.

Picture Credits

Cover: Edward Hicks, *Penn's Treaty with the Indians*, Gilcrease Museum, Tulsa; p. 5: Library of Congress; p. 6: from André Thevet, *Les Singularitéz de la France Antarctique, autrement nommé Amérique*, engraved by Silvanus Antoianus, printed by Christopher Plantin, Antwerp, 1557; pp. 6–7: Library of Congress; p. 7: National Portrait Gallery, Smithsonian Institution; p. 9: Metropolitan Museum of Art, Harry Brisbane Dick Fund, 1959; p. 10 (top): Picture Collection, New York Public Library; p. 10 (bottom): Istitutuo Arti Grafiche, Bergamo; p. 11: Picture Collection, New York Public Library; pp. 14, 15 (lower right): Library of Congress; pp. 16, 17: Picture Collection, New York Public Librar;y; p. 18: Library of Congress; p. 19: from Arnold Montanus, *De Nieuwe En Onbekende Weereld, America*, Amsterdam, 1671; p. 20 (upper left): from André Thevet, *Les Singularitéz de la France Antarctique, autrement nommé Amérique*, engraved by Silvanus Antoianus, printed by Christopher Plantin, Antwerp, 1557; p. 20 (bottom): title page from *The New Life of Virginea,* printed by Felix Kyngston for William Welby, London, 1612; pp. 21, 23–26, 28–31, 33: Library of Congress; p. 35: National Portrait Gallery, Smithsonian Institution; p. 36: Library of Congress; p. 37 (upper left), 38 (upper left): New-York Historical Society; p. 38 (bottom): Library of Congress; p. 40 (top): Picture Collection, New York Public Library; p. 41 (bottom): Library of Congress; p. 42: Rare Books and Manuscripts, New York Public Library; p. 43: British Museum; p. 44: Rare Books and Manuscripts, New York Public Library; p. 45 (top): Colonial Williamsburg Foundation; p. 45 (bottom): Joe Flying Horse, "Around the Rez," *The Lakota Times* (now called *Indian Country Today*), September 1989; pp. 46, 47: L:ibrary of Congress; pp. 48-49: Colonial Williamsburg Foundation; p. 51 (left): Picture Collection, New York Public Library; p. 51 (right): Library of Congress; p. 52: Dutch School, from *Harper's Weekly*, March 9, 1895; p. 54: Library of Congress; p. 55 (top left): A. R. Waud, *Trapping Wild Turkeys*, from *Harper's Weekly*, November 28, 1868; p. 55 (bottom): Picture Collection, New York Public Library; p. 56: Library of Congress; p. 57: American Antiquarian Society; p. 58: Massachusetts Archives; p. 59: Library of Congress; p. 60: Picture Collection, New York Public Library; p. 61 (upper left): British Library; p. 61 (bottom): Library of Congress; p. 62: (top): British Museum; p. 62 (bottom left): courtesy, Winterthur Museum; p. 62 (bottom right), 63: Library of Congress; p. 65: Historical Society of Pennsylvania; pp. 66, 67: Library of Congress; p. 68: Picture Collection, New York Public Library; p. 69: Pierpont Morgan Library; p. 70 (upper): from Samuel Seyer, *Memoirs Historical & Topographical of Bristol*, New York Public Library; p. 70 (bottom): Quaker Meeting, Museum of Fine Arts, Boston; p. 71: Library of Congress; p. 72 (top): Picture Collection, New York Public Library; p. 72 (bottom): from Matthew Hopkins, *Discoverie of Witches,* New York Public Library; p. 73: British Library; p. 74: Library of Congress; p. 75 (upper): Massachusetts Historical Society; pp. 76, 77 (upper left): Library of Congress; p. 79: courtesy John Carter Brown Library at Brown University; p. 80: Library of Congress; pp. 81, 82: Library of Congress; p. 83 (bottom): *Hartford Courant*; p. 84: from Fr. Pablo Beaumont, *Cronica de Mechoacan*, Rare Books and Manuscripts, New York Public Library; p. 86 (top): courtesy, Museum of New Mexico, Neg. No. 133131; p. 86 (bottom): from Fr. Pablo Beaumont, *Cronica de Mechoacan*, Rare Books and Manuscripts, New York Public Library; p. 87: courtesy, Museum of New Mexico, Neg. No. 11409; p. 88: Library of Congress; p. 89: Picture Collection, New York Public Library; p. 91: Library of Congress; p. 92 (upper): New York Historical Society; p. 92 (bottom): Library of Congress; p. 93: *New Amsterdam, Prototype View, 1650-53*, J. Clarence Davies Collection; Museum of the City of New York; p. 93 (insets): Library of Congress; p. 94: Library of Congress; pp. 97, 98: Library of Congress; p. 99: Radio Times Hulton Picture Library; p. 100: from H.D. Traill, *Social England*, New York Public Library; p. 101 (upper left): Library of Congress; p. 101 (bottom right): title page from Gabriel Thomas, *Pensilvania and West-New-Jersey*, published by A. Baldwin, London, 1698; p. 102: Picture Collection, New York Public Library; p. 102 (bottom): Library of Congress; p. 103: Map Division, New York Public Library; p. 104: Edward Hicks, *Penn's Treaty with the Indians*, Gilcrease Museum, Tulsa; p. 105: Picture Collection, New York Public Library; p. 106 (center and bottom right): Library of Congress; p. 106 (top left): Rare Books and Manuscripts, New York Public Library; p. 106 (top right and bottom left): Historical Society of Pennsylvania; p. 107: Harvard University; p. 108: from Holley, *The Life of Mr. Benjamin Franklin*, 1848, Benjamin Franklin Collection, Yale University Library; p. 110: Library of Congress; p. 111 (upper): Picture Collection, New York Public Library; p. 111 (bottom): Library of Congress; p. 112 (two upper left): Picture Collection, New York Public Library; p. 112 (top right): Baltimore Museum of Art; p. 113 (top): Picture Collection, New York Public Library; p. 113 (bottom): Library of Congress; p. 115: American Antiquarian Society; p. 116 (top): Dementi/Foster Sudios, Richmond, Virginia; p. 116 (inset): Picture Collection, New York Public Library; p. 117 (top): New York Public Library; p. 117 (bottom): Justus Englehardt Huhn, *Eleanor Darnall*, circa 1710, Maryland Historical Society; p. 118 (top): Library of Congress; p. 118 (bottom): Colonial Williamsburg Foundation; p. 120 (right): sampler by Mary Batchelder, 1773, Cooper-Hewitt Museum, Smithsonian Insitution; p. 121: Library of Congress; p. 122: Colonial Williamsburg Foundation; p. 123: Muscarelle Museum of Art; pp. 124, 125: Colonial Williamsburg Foundation; p. 126: Library of Congress; p. 127 (top): Library of Congress; p. 128: *The Old Plantation*, Abby Aldrich Rockefeller Folk Art Center, Williamsburg, Virginia; p. 129 (bottom): Charleston Year Book, 1884; p. 130: Louis Manigault after George Roupell, *Mr. Peter Manigault and His Friends*, watercolor and ink on paper, Gibbes Museum of Art/Carolina Art Association; p. 131 (top): *Harper's Weekly,* January 5, 1867; p. 131 (bottom): Charleston Museum; p. 132 (top): New-York Historical Society; p. 132 (bottom): from Johnson, *Lives of High-Waymen and Pirates*, 1736, New York Public Library; p. 133 (top two): from Johnson, *Historie der Engelsche Zee-roovers*, 1725, Harvard College Library, Harry Elkins Widener Collection; p. 133 (bottom): from Johnson, *History of Pirates*, 1724, New York Public Library; p. 134: Library of Congress; p. 135 (top): Picture Collection, New York Public Library; p. 135 (bottom): *Hartford Courant*; p. 136 (top): Schomburg Center for Research in Black Culture, New York Public Library; p. 136 (bottom): New York Public Library; pp. 137, 138: Library of Congress; p. 141: Picture Collection, New York Public Library; p. 142: *Nieu Amsterdam*, Stokes Collection, New York Public Library; p. 143 (top right): Guildhall Library, London; pp. 143, 144: Library of Congress; p. 145: courtesy, Winterthur Museum; p. 146 (top): Museum of Science and Industry, Chicago; p. 146 (bottom): Cincinnati Historical Society; p. 147: Library of Congress; p. 148: George Tattersall, *Highways and Byeways of the Forest*, M. and M. Karolik Collection, Museum of Fine Arts, Boston; p. 149: Picture Collection, New York Public Library; pp. 150, 152: Library of Congress; p. 153: Washington/Custis/Lee Collection, Washington and Lee University, Lexington, VA; p. 156: from Johnson, *The History of Pirates*, 1724, New York Public Library.

The pirate Blackbeard

Index

A Note from the Author

My editor, whose name is Tamara Glenny, says I have to write about myself on this page. And editors are like teachers, you do what they tell you to do. But, since I already told you all the important things about me in Book 1 of *A History of US*, I will tell you a secret in this book. My real name is Natalie. Natalie Joy Frisch Hakim.

Whew! I haven't told anyone that in a long time, except for my granddaughter, whose name is also Natalie. I wrote this book for her.

Now that you know two things about me, I will also tell you that I live in Virginia with my husband Sam, and a dog named Claire. I live on an old sand dune not far from Cape Henry, the very spot where John Smith and the first English settlers landed. (Yes, there is a house on the sand dune. It is made of wood and has a bridge inside.) In my neighborhood the streets are named Susan Constant, Discovery, and Godspeed, and, if you've read this book carefully, you know why that pleases me.

Tamara says I have to tell you more. Here is more: I graduated from Rutland High School (in Rutland, Vermont), and got a B.A. from Smith College (in Northampton, Massachusetts), and an M.Ed. from Goucher College (in Baltimore, Maryland). I've taken courses at lots of other colleges, too. I love to go to school and I love to read.

Some people tell me they hate to read and they don't like school much. I'm sure that is because they haven't found good books or the right teachers. My advice is to keep looking; there is nothing as exciting as learning things.

I'm almost finished with this story. I have three children who are now grown-ups themselves. Ellen is beautiful and does an astounding number of things and does them all well. If you see my son Jeffrey you may get a sore neck looking up at his head—the top of it is 6 feet 4 inches from the ground. Jeff is a mathematics professor and a mighty nice one. And then there is Danny, who has curly brown hair and is an artist and a writer, too. In the next book you will learn the name of my grandson.